# Digital Counter Handbook

by Louis E. Frenzel, Jr.

Howard W. Sams & Co., Inc.
4300 WEST 62ND ST. INDIANAPOLIS, INDIANA 46268 USA

Copyright ©1981 by Louis E. Frenzel, Jr.

FIRST EDITION
FIRST PRINTING—1981

All rights reserved. No part of this book shall be reproduced, stored in a retrieval system, or transmitted by any means, electronic, mechanical, photocopying, recording, or otherwise, without written permission from the publisher. No patent liability is assumed with respect to the use of the information contained herein. While every precaution has been taken in the preparation of this book, the publisher assumes no responsibility for errors or omissions. Neither is any liability assumed for damages resulting from the use of the information contained herein.

International Standard Book Number: 0-672-21758-9
Library of Congress Catalog Card Number: 80-52938

*Printed in the United States of America.*

# Preface

The term "digital counter" can be used to describe a variety of different electronic circuits and equipment. For example, in one context a digital counter is a circuit made up of cascaded flip-flops that accumulate input pulses and store them as a binary or bcd number. Such digital counters are the heart of most digital designs.

The term digital counter is also applied to special-purpose instruments for industrial measurement and control. Both electronic and electromechanical counters are used to tally events, count objects, time operations, and perform unit conversions.

A digital counter can also be a piece of test equipment. Digital counters of this kind are used in laboratories for research, development, and design; they are found in manufacturing operations for testing; and they are used by most electronic servicemen for troubleshooting, servicing, measurement, and alignment. This book covers all three types of digital counters, but the emphasis is on digital-counter test instruments.

Despite its growing numbers, the digital counter is still a poor third in terms of popularity behind the common multimeter and the oscilloscope. It is perhaps the most underrated and underused test instrument. Its basic simplicity in function and operation tends to mask its true value. It is hoped that this book will bring to light the true versatility of the digital counter so that it will no longer be taken for granted.

Electronic counters are no longer large, expensive, and exotic instruments used in a few select laboratories. Thanks to modern semiconductor technology, powerful and sophisticated digital counters are now available at reasonable prices, and virtually everyone can afford a good basic counter.

The main purpose of this book is to explain how digital counters work and to show how they are used. The book helps you to understand counter specifications and to apply the counter to a wide variety of electronic measurements.

Counters are most often used for measuring the frequency of electronic signals. Time intervals can also be measured. Electrical impulses, both periodic and aperiodic, can be counted, tallied, or accumulated with a counter. In all cases, the signal to be measured is applied to the counter input, and the time, frequency, or tally is displayed on numerical readouts.

Because the time and frequency characteristics of electronic signals are so important, electronic counters find application in virtually every electronic field. They are widely used in am, fm, and tv broadcasting. They are also used in mobile radio and other communications work. The counter is used as a basic test instrument in troubleshooting, servicing, and adjusting almost all types of electronic equipment. Counters are frequently needed in electronic circuit design. Electronic counters are your eyes and ears for all types of time and frequency measurements. They provide an accuracy unmatched in any other type of test instrument.

Despite the versatility and growing popularity of counters, little practical applications information is available. The only apparent reference sources are maufacturers' literature, data sheets, and manuals. A few general textbooks on test equipment are available, but their coverage of electronic counters is limited. This book is intended to fill this void of practical information on the operation, specifications, and applications of digital counters.

This book is written for engineers, technicians, scientists, hobbyists, students, and others looking for detailed technical information on counters. It is a down-to-earth working guide you can keep near your workbench for handy reference. I think this book will stimulate your interest in electronic counters. I hope that by the time you complete it, you will have a thorough appreciation of the power and usefulness of modern electronic counters and some practical knowledge of how to get the most from your counter investment.

Lou Frenzel

# Contents

### CHAPTER 1

INTRODUCTION TO DIGITAL COUNTERS............................................. 7
  Counter Circuits—The Electronic-Counter Test Instrument—Industrial Counters

### CHAPTER 2

THE COUNTER TEST INSTRUMENT.................................................. 20
  Counter Components—Counter Operational Modes

### CHAPTER 3

COUNTER SPECIFICATIONS AND ERROR SOURCES................................. 55
  Performance Specifications—Error Sources

### CHAPTER 4

HIGH- AND LOW-FREQUENCY COUNTERS........................................... 75
  Prescaling—Heterodyne Down Conversion—Transfer-Oscillator Down Conversion—The Harmonic-Heterodyne Converter—Specifications of High-Frequency Counters—Low-Frequency Counters

## CHAPTER 5

COUNTER ACCESSORIES................................................................. 108
    Probes, Cables, and Connectors—Preamplifiers—Prescalers—Interfaces

## CHAPTER 6

COUNTER MEASUREMENT APPLICATIONS........................................ 132
    Basic Measurement Principles—Frequency Measurement—Time Measurements—Complex Signal Measurements

## CHAPTER 7

SPECIAL COUNTERS........................................................................ 156
    Computing Counters—Industrial Counters

## CHAPTER 8

TYPICAL DIGITAL COUNTERS......................................................... 178
    Handheld LSI Counter—LSI Communications Counter—General-Purpose Bench Counter—Universal Counter-Timer

## CHAPTER 9

COUNTER CIRCUITS AND APPLICATIONS........................................ 217
    MSI Counters—Synchronous Versus Asynchronous—Reset and Preset Functions—Up/Down Counters—Counters as Dividers—Applications of Binary and BCD Counters—Timers—Rate Multiplier—LSI Counters—Calculators as Counters—Microprocessors as Counters—Microprocessor Peripheral Counter/Timers

INDEX............................................................................................. 259

CHAPTER 1

# Introduction to Digital Counters

The term "digital counter" describes a variety of different electronic circuits and equipment. For example, a digital counter is a widely used digital circuit, made up of cascaded flip-flops, that accumulates input pulses and stores them as a binary or bcd number. Such digital counters are the cornerstone of most digital circuit designs.

A digital counter is also a piece of test equipment. Digital counters can be found on most electronic workbenches. They are used in laboratories for research, development, and design; they are found in manufacturing operations for testing; and they are used by most electronic servicemen for troubleshooting, servicing, measurement, and alignment.

Digital counters are also special-purpose instruments for industrial measurement and control. Both electronic and electromechanical counters are used to tally events, count objects, time operations, and perform measurement conversions.

This book covers all three types of digital counters. However, the emphasis is on digital-counter test instruments. Their operation and applications are thoroughly discussed. Industrial counters are usually simpler, special-purpose versions of the counter test instrument. These counters and their applications are also covered. Finally, digital counter circuits used in both the test-instrument counters and industrial counters are explained as they apply in these applications. The result is complete coverage of the broad spectrum of digital counters available today.

## COUNTER CIRCUITS

One of the most frequently used digital circuits is a counter. A counter is made up by cascading flip-flops. A flip-flop is a two-state digital logic circuit that is used to store binary data. A review of flip-flop operation is given in Reference I.

By connecting several flip-flops together, one driving another as shown in Fig. 1-1A, a counter is created. Called a *binary counter*, this

---

## REFERENCE I

### Review of Flip-Flop Operation

A flip-flop is a two-state, or binary, logic circuit used to store one bit of binary data. If the flip-flop is in its *set* state, it is said to be storing a binary 1. If the flip-flop is in its *reset* state, it is said to be storing a binary 0.

Fig. I-1 shows the logic symbol used to represent a flip-flop. Various input and output lines are used to control and monitor the state of the flip-flop.

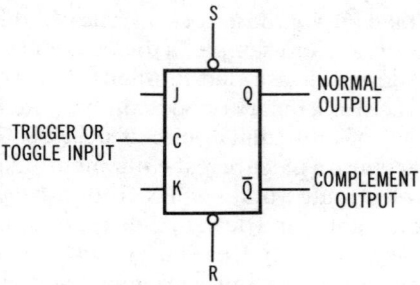

Fig. I-1. Logic symbol for JK flip-flop.

The output lines are usually designated by a letter of the alphabet, such as Q. Two outputs, *normal* and *complement*, are available. To determine the state of the flip-flop, the normal output is monitored. If Q = 1, then the flip-flop is set. If Q = 0, then the flip-flop is reset. The complement output, $\overline{Q}$, is always the opposite (complement) of the normal Q output.

The S (set) and R (reset) input lines are used to preset the flip-flop to a given state. Usually, applying a binary 0 logic level to R resets the flip-flop and applying a 0 level to S sets the flip-flop.

The C (clock) input is used to toggle, or complement, the flip-flop. On the trailing edge (the transition from binary 1 to binary 0) of the pulse applied to C, the flip-flop will change states as the waveforms in Fig. I-2 indicate. Note that the frequency of the output is one-half the frequency of the input, making the flip-flop a 2-to-1 frequency divider. If the C input signal is a 50–kHz square wave, the Q output is a 25-kHz square wave.

The toggling, or complementing, action of the C input occurs if both the

*Continued on next page.*

REFERENCE I *(cont.)*

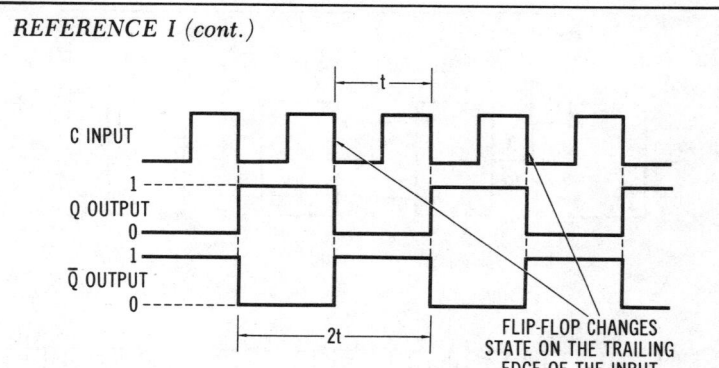

Fig. I-2. Flip-flop waveforms.

J and K inputs are binary 1. These inputs can also control the state of the flip-flop. By applying the appropriate logic levels to the J and K inputs, the flip-flop can be set or reset, but only upon the occurrence of a pulse trailing edge at the C input. The table below shows the effect on the state of the flip-flop for various J and K inputs.

| INPUTS | | OUTPUT | |
|---|---|---|---|
| J | K | Q | $Q_{t+1}$ |
| 0 | 0 | X | X |
| 0 | 1 | X | 0 |
| 1 | 0 | X | 1 |
| 1 | 1 | X | $\overline{X}$ |

In this table, Q is the state of the flip-flop prior to a clock pulse. State X can be either a 0 or a 1. Column $Q_{t+1}$ shows the state of the flip-flop after a clock-pulse trailing edge occurs with the given J and K input states.

circuit is capable of counting input pulses applied to the circuit and storing the number of pulses as a binary number.

For example, assume that all of the flip-flops in Fig. 1-1A are initially reset so that the DCBA outputs are 0000. This is the binary number representing the decimal value zero. Now assume that the input pulses shown in Fig. 1-1B are applied to flip-flop A. Each time the trailing edge (1-to-0 transition) of an input pulse occurs, flip-flop A changes state. Flip-flop A drives B, B drives C, and C drives D so that each flip-flop changes state when it receives a trailing-edge input at its clock (C) terminal. By observing the normal flip-flop outputs, the waveforms shown in Fig. 1-1B are obtained. Assume that the trailing edge of input pulse number 5 has just occurred. Referring to the waveforms, you can

9

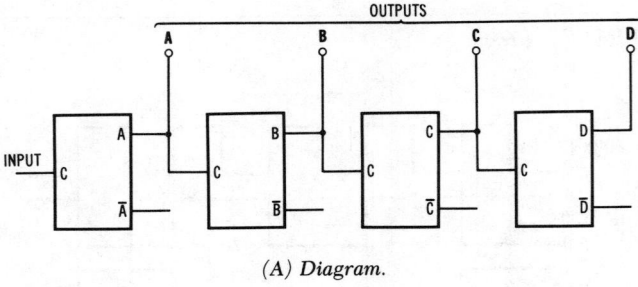

*(A) Diagram.*

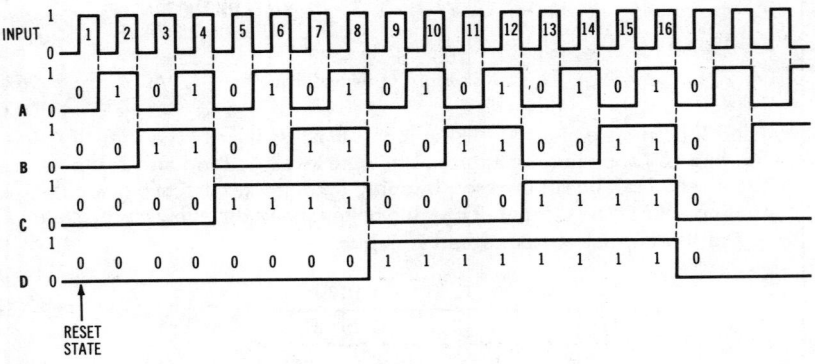

*(B) Waveforms.*

Fig. 1-1. Four-stage binary counter consisting of cascaded flip-flops.

see that flip-flop A is set (1), B is reset (0), C is set (1) and D is reset (0). Reading these flip-flop states in the order DCBA, we get the binary number 0101, or decimal 5. This clearly shows that the counter indicates that five input pulses have occurred. If outputs DCBA were connected to indicators such as LEDs, the state of the counter and the number of input pulses could be observed.

The four-state counter in Fig. 1-1 is capable of counting up to 15 pulses. After the fifteenth input pulse, the counter output DCBA reads 1111, or the binary equivalent of the decimal number 15. After the sixteenth pulse, all flip-flops reset. If a fifth flip-flop, E, were added, when the D flip-flop reset on the sixteenth pulse, the E flip-flop would be set, thus recording the number EDCBA = 10000, or the binary number for a decimal 16.

The maximum number that a counter can store is determined by the expression

$$M = 2^N - 1$$

where,
M is the maximum number,
N is the number of flip-flops.

With four flip-flops, the maximum number is

$$M = 2^4 - 1 = 16 - 1 = 15$$

as indicated earlier. Any number can be stored if a sufficient number of flip-flops are cascaded. Cascading eight flip-flops would allow you to record numbers up to $2^8 - 1 = 256 - 1 = 255$.

Binary numbers are difficult to work with in that you must convert from binary to decimal to determine the count. This is time-consuming and inconvenient. For example, suppose the LED display on an eight-bit counter read 10010101. How many input pulses occurred? Converting binary to decimal, you get the number 149. To overcome this binary-to-decimal conversion difficulty, the *bcd* counter was developed. The abbreviation bcd means *binary-coded decimal*, which is a special four-bit code used to represent the ten digits 0 through 9. See Reference II. The bcd counter is designed to count in this sequence.

One kind of bcd counter circuit is shown in Fig. 1-2A. Note that four cascaded flip-flops are used. The extra AND gate in the circuit "tricks" it into counting in the bcd code. You can check the waveforms in Fig. 1-2B

### REFERENCE II

### Binary Coded Decimal Numbers

| Decimal Number | BCD Code | | | |
|---|---|---|---|---|
| | D | C | B | A |
| 0 | 0 | 0 | 0 | 0 |
| 1 | 0 | 0 | 0 | 1 |
| 2 | 0 | 0 | 1 | 0 |
| 3 | 0 | 0 | 1 | 1 |
| 4 | 0 | 1 | 0 | 0 |
| 5 | 0 | 1 | 0 | 1 |
| 6 | 0 | 1 | 1 | 0 |
| 7 | 0 | 1 | 1 | 1 |
| 8 | 1 | 0 | 0 | 0 |
| 9 | 1 | 0 | 0 | 1 |

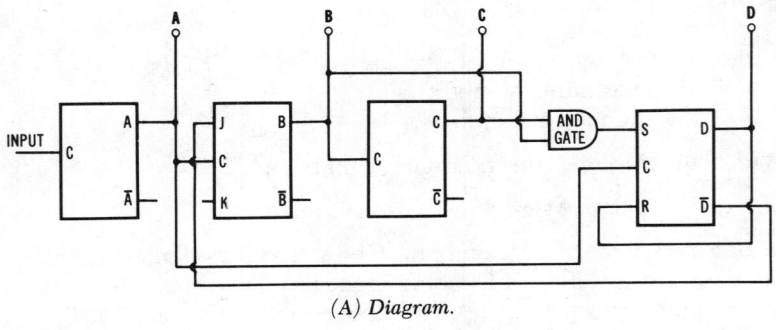

(A) Diagram.

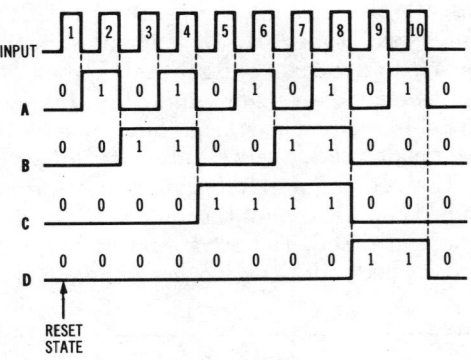

(B) Waveforms.

Fig. 1-2. A binary-coded-decimal (bcd) counter.

against the code in Reference II to verify that they are the same. Notice that after the ninth input pulse, the counter code DCBA is 1001, or decimal 9. After the tenth input pulse, the counter recycles to zero (0000).

What good is a counter that only counts up to 9? By itself it has only limited value, but if we cascade bcd counters, we can represent any decimal value. Refer to Fig. 1-3. Here three bcd counters are connected so that one triggers another in sequence. For every ten input pulses, the first counter cycles from 0 through 9, then on the tenth pulse resets to zero. Every time the first counter cycles from 9 back to 0, it triggers the second counter. As you can see, the input counter counts by units, and the second counter counts by tens. Once the second counter counts ten pulses, it triggers the third counter, indicating one hundred pulses. If a fourth bcd counter were added, it could count by thousands, and so on.

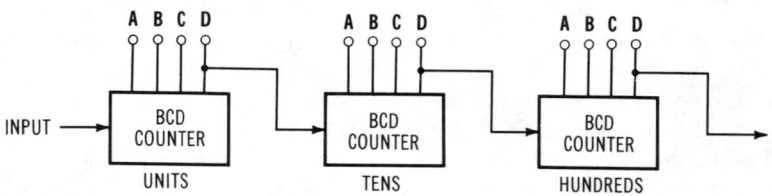

Fig. 1-3. A cascaded bcd counter.

By cascading the appropriate number of bcd counters, any decimal number can be recorded.

To determine the content of a bcd counter, you observe the flip-flop outputs. You simply look at the DCBA outputs of each in sequence, keeping in mind that the least significant digit counter (units) is the input counter. For example, if the outputs of the counter in Fig. 1-3 were units DCBA = 0100, tens DCBA = 1001, and hundreds DCBA = 0011, the decimal number in the counter would be 394. Simply convert each four-bit code into its decimal equivalent by referring to Reference II. Therefore, if each bcd counter is supplied with LED output indicators, you could determine the content by observing the output states. Since each bcd counter in Fig. 1-3 can count up to 9, the maximum count that can be represented by the three counters is 999. Higher values can be accommodated by adding additional bcd counters. Because bcd counters count by ten, they are called *decade counters*.

The bcd, or decade, counter is the heart of most modern digital counters. Whether an instrument is a bench counter for measuring frequency or an industrial counter for tallying the number of widgets coming off a production line, the main circuit element is a decade counter. Such counters are made easier to use by adding decimal-readout display units instead of binary indicator lights. The bcd outputs of each decade counter are fed to a bcd-to-decimal decoder logic circuit that in turn drives a decimal number such as the widely used 7-segment LED display. The output result can be quickly and easily read by anyone.

Most counters are made up of integrated-circuit flip-flops. These can be interconnected in a variety of ways to form counters of any length or configuration (*i.e.*, binary or bcd). However, the integrated-circuit manufacturers have made it even easier for us to use counters. They have developed a variety of MSI and LSI single-chip counters. Four- and eight-bit binary counters and bcd counters containing as many as eight cascaded bcd stages are also available. With such devices, counters can be widely applied to many applications at very low cost.

# THE ELECTRONIC-COUNTER TEST INSTRUMENT

The digital counter is one of the most popular electronic test instruments in use today. Next to the common multimeter and the oscilloscope, the digital counter has become one of the more valuable pieces of test equipment. Yet, despite its popularity, it is still a poor third to multimeters and oscilloscopes. It is perhaps the most underrated, under-used, and misunderstood piece of test equipment. Its basic simplicity in function and operation tends to mask its true value and flexibility. It is hoped that this book will bring to light the true versatility of the digital counter so that it will no longer be taken for granted.

A digital counter is a general-purpose test instrument used for making time and frequency measurements. Counters are most often used for measuring the frequency of electronic signals. Therefore, they are often referred to as frequency meters or frequency counters. Time intervals and period can also be measured. Electrical impulses, both periodic and aperiodic, can be counted, tallied, or accumulated with a counter. In all cases, the signal to be measured is applied to the counter input, and the time, frequency, or tally is displayed on numerical readouts. Fig. 1-4 shows a photo of three general-purpose digital-counter instruments.

The first generation of universal, general-purpose counters was introduced about 25 years ago. These instruments were implemented with vacuum tubes and were large and expensive. While these counters had limited frequency range and versatility, they brought a new dimension of measuring capability to the laboratory. It was possible at last to obtain reliable and accurate measurement of frequency and time on a practical everyday basis.

In the early sixties, the second generation of counter test instruments was introduced. Thanks to transistors, these units were smaller, consumed a lot less power, and were lower in cost. While their measurement capability was not significantly greater than that of their vacuum-tube ancestors, these counters became far more widely used.

In the late sixties, MSI integrated circuits were introduced. These devices made a major impact on counter design, and a third generation of counters was born. Not only were smaller, less expensive counters possible, but also such counters had added versatility and improved performance.

A further development in integrated circuits, large scale integration (LSI), has created yet a fourth generation of digital counters. These counters are smaller and more powerful than ever. Virtually all counter functions are contained on a single LSI chip. As a result, cost and size have continued to decline.

The availability of microprocessors has also impacted counter design. By combining an MOS LSI counter chip with a microprocessor, a very

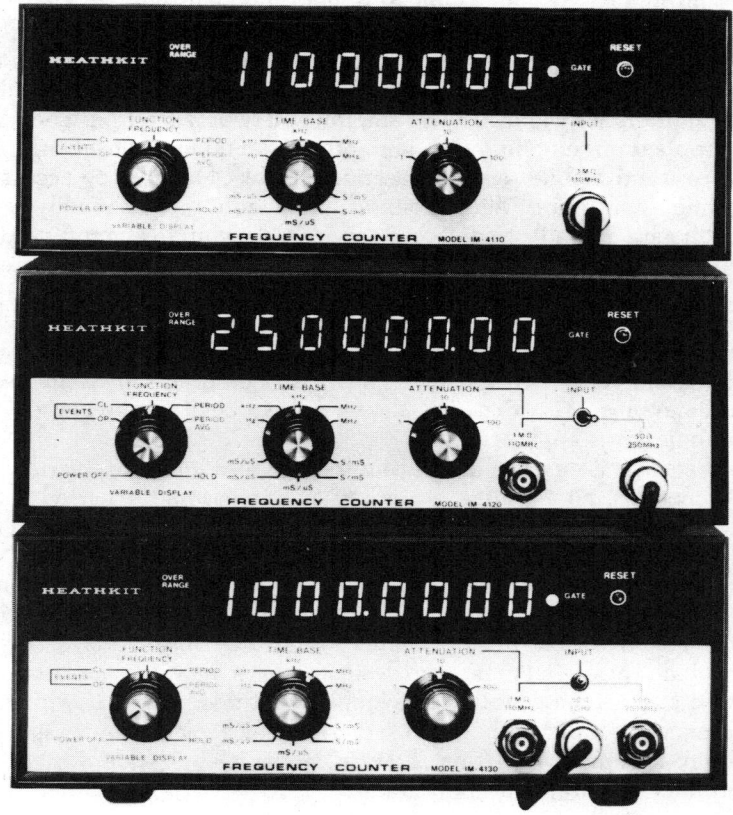

Courtesy Heath Co.

Fig. 1-4. Typical digital-counter test instruments.

powerful and versatile general-purpose test instrument results. Many sophisticated measurements can be made. In addition, automatic self-checking operation is possible.

Thanks to modern semiconductor technology, powerful and sophisticated digital counters are now available at extremely low prices. Electronic counters are no longer large, expensive, and exotic instruments for use in a few selected laboratory situations. Today, many excellent counters are available for less than $100.

Because the timing and frequency characteristics of electronic signals are so important, electronic counters find application in virtually every electronic field. Counters are used as a basic test instrument in design, troubleshooting, servicing, and adjusting almost all other types of electronic equipment. Electronic counters are your eyes and ears for all types of time and frequency measurement functions with an accuracy unmatched in any other type of test instrument.

The main use of most digital counters is frequency measurement. Some of the smaller, lower-cost units in fact measure frequency only. Such counters are widely used in am, fm, and tv broadcasting. They are also applied in mobile radio and other communications work. In broadcast and mobile radio applications, the FCC insists on accurate frequency monitoring. With a counter, this requirement is readily met. Counters are available to measure any frequency up to about 50 GHz. Even subhertz measurements can be made. With high measurement accuracy and the convenience of direct decimal frequency display, the counter adds an element of speed, simplicity, and convenience to measurements that would otherwise be difficult to make. Counters make earlier frequency-measuring devices look crude in comparison (see Reference III). Fig. 1-5 shows a frequency counter used in communications applications.

Some general-purpose bench counters can also measure a variety of time functions. For example, a counter might measure the period (time for one cycle) of an input signal. Instead of displaying a frequency of 1000 Hz, the counter would display a period of 1 millisecond (t = 1/f = 1/1000 = 0.001 second, or 1 millisecond). Many different pulse measurements can also be made with a counter. Pulse width, pulse spacing, and rise time are some of the measurements that can be made (Fig. 1-6).

If you do any digital rf or audio design or servicing work, you are sure to want and need a counter. The emphasis of this book is how digital counters work and how they are used. The book explains how to specify and buy a counter and then how to apply it to a wide variety of electronic-measurement applications.

Courtesy Davis Electronics

Fig. 1-5. A low-cost frequency counter.

# REFERENCE III

## Early Frequency-Measuring Devices

It has been within only the past 20 years that low-cost counters have been available to make accurate frequency measurements. What was used before counters? Wavemeters and heterodyne frequency meters were two of the popular instruments now almost completely replaced by modern digital counters.

### WAVEMETERS

A wavemeter is a simple instrument consisting of a coil, a precision calibrated variable capacitor, and an ammeter, all connected as a series resonant circuit. The coil (inductor) is inductively coupled to the circuit, such as the output tank circuit of a radio transmitter, whose frequency is to be measured (Fig. III-1). The transmitter induces a voltage into the wavemeter circuit, and the resulting current is indicated on the ammeter. The variable capacitor is tuned for a maximum reading on the ammeter. At this point, the circuit is resonant (inductive and capacitive reactance equal) at the output frequency of the transmitter. The frequency is then read from the calibrated capacitor dial.

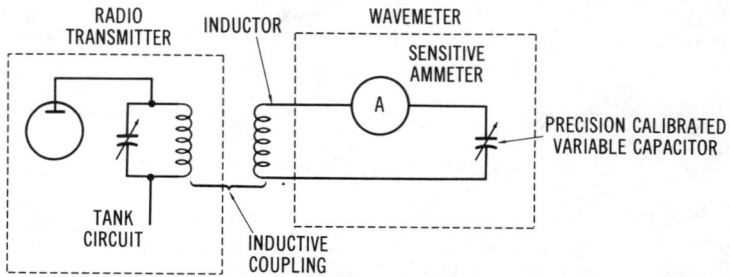

Fig. III-1. Operation of wavemeter.

### HETERODYNE FREQUENCY METER

A heterodyne frequency meter measures an unknown frequency ($f_x$) by comparing it to a locally generated signal of the same frequency. The frequency meter contains a precision, calibrated variable frequency oscillator (vfo) that generates an output frequency, $f_o$. The vfo is calibrated by matching its frequency to a highly accurate and stable internal crystal oscillator.

An unknown frequency is measured by applying it to a mixer circuit in the frequency meter along with the vfo signal (Fig. III-2). Frequencies $f_x$

*Continued on next page.*

REFERENCE III (cont.)

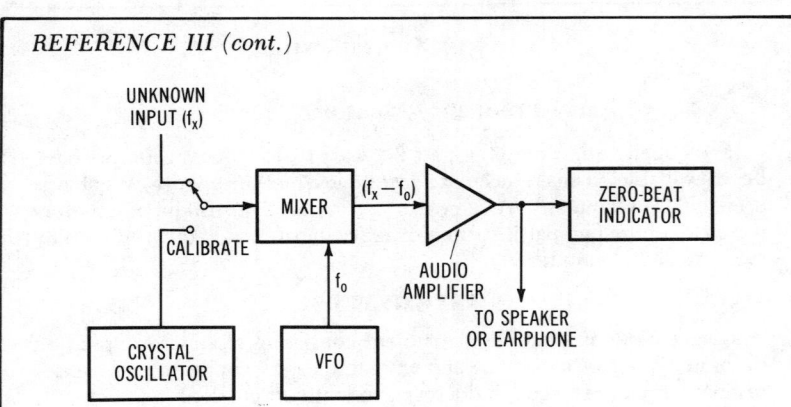

Fig. III-2. Operation of heterodyne frequency meter.

and $f_o$ are mixed, or heterodyned. The mixer generates the sum ($f_x + f_o$) and difference ($f_x - f_o$) outputs. It is the difference output that is used to compare the two input signals.

The vfo is tuned until $f_o$ is nearly the same as $f_x$. This is indicated by an audio signal that is the difference between the two inputs. The audio signal can be heard in a speaker or earphone connected to the audio amplifier output. The vfo is tuned for lowest or zero frequency. This is called *zero beat*, and at this point $f_x = f_o$ since $f_x - f_o = 0$. At zero beat, the unknown frequency is read from the calibrated vfo dial.

## INDUSTRIAL COUNTERS

Industrial counters are a special class of counters used for making a variety of time and frequency measurements in industrial control and monitoring applications. These counters fall somewhere between simple counter circuits and deluxe universal counter test instruments.

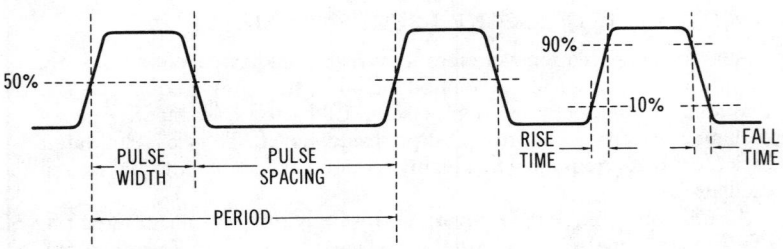

Fig. 1-6. Some characteristics of pulse waveforms that can be measured with a counter.

Courtesy Electro-Numerics Inc.

**Fig. 1-7. A preset industrial counter.**

Industrial counters usually consist of a chain of bcd counters with decimal displays and a variety of controls. These counters are normally dedicated to a single function. For example, an industrial counter might be used to count the number of boxes being moved past a certain point on a conveyer belt. Other counters, when combined with appropriate input transducers, are used to monitor motor speed in rpm (tachometer), measure the length of a sheet of aluminum, or count the number of welds made by an automatic welder and then automatically shut it off after a specific number has been reached. A special form of industrial counter, more accurately called a timer, is widely used to measure time in hours, minutes, and seconds and initiate or stop some function at predetermined times.

A typical industrial counter is shown in Fig. 1-7. These small, low-cost units can be customized to a variety of applications in plants and factories. They give the precise readout and control of quantity so important in achieving efficiency, economy, and safety in manufacturing. A high percentage of industrial counters are fully electronic, but mechanical, electromechanical, and pneumatic counters are also available for special applications or severe environments. Industrial counters are extremely simple, but their simplicity belies their usefulness.

CHAPTER 2

# The Counter Test Instrument

If you work in electronics, you are sure to encounter a digital counter sooner or later. As an engineer, you may need a digital counter to aid you in some design job. As a technician, you will no doubt find a counter to be a real benefit in your testing, repair, and service work. If you are a scientist, you may need the high-precision time and frequency measuring capability of a counter in one of your experiments. Even as a hobbyist, you may want or need an electronic counter to help you measure the frequency of your ham transmitter or aid you in making one of your experimental circuits work. Regardless of the specific application, your ability to use the counter properly and get the most value from it will come if you know counter operation, specifications, and characteristics. The purpose of this and the following chapters is to give you a good understanding of how digital counters operate and how they are used. With this background, you will be able to apply the counter with confidence.

Fig. 2-1 shows two digital-counter test instruments. These counters are typical of those general-purpose instruments used in laboratory development and service work. The internal circuitry discussed in this chapter is typical of what you will find in counters of this type.

## COUNTER COMPONENTS

A general block diagram of a digital-counter test instrument is shown in Fig. 2-2. Virtually all digital counters are made up of six basic components. These are the input circuit, the gate, the decimal counter, the display, the control circuits, and the time base. These circuits in various combinations permit the counter to make a variety of time and frequency measurements. Both the simplest and the most sophisticated

Courtesy John Fluke Mfg. Co., Inc.

(A) Counter-timer.

Courtesy Dynascan Corp.

(B) Universal frequency counter.

Fig. 2-1. Typical counter instruments.

counters contain these six basic elements. However, the complexity, sophistication, and precision with which these circuits operate determine the capabilities, characteristics, and limitations of the

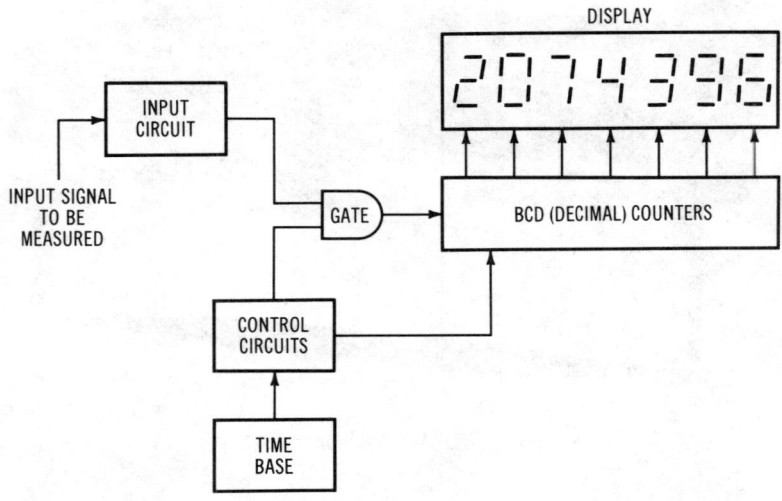

Fig. 2-2. Block diagram of a digital-counter test instrument.

counter. In this section, we will discuss each of the circuits individually, thereby defining it and explaining its function. In the next section, we will see how these circuits work together to perform various time and frequency measurements.

**The Input Circuit**

The input signal to be measured is applied to the input circuit. The input circuit is primarily a signal conditioner. It takes the input signal and converts it into a voltage with the proper amplitude and shape for reliable operation of the digital logic circuitry in the counter. The input circuit is probably the most critical in the counter because it must deal with a wide range of signal characteristics including amplitude, frequency, and shape.

The input circuit generally consists of five basic parts, as illustrated in Fig. 2-3. It consists of an ac/dc coupling circuit, an input attenuator, a voltage limiter, an amplifier, and a trigger circuit.

The ac/dc coupling circuit allows optimum measurement of both ac and dc signals. For ac signals, the coupling circuit consists primarily of a capacitor. For dc operation, or direct coupling, the capacitor is eliminated, and the signal is connected directly to the input attenuator. The dc input is used primarily for pulse and logic-signal measurements, whereas the ac input is used primarily with sine-wave ac signals or with signals that have a dc component which must be eliminated. A front-panel switch is usually provided to allow the user to select the desired input coupling.

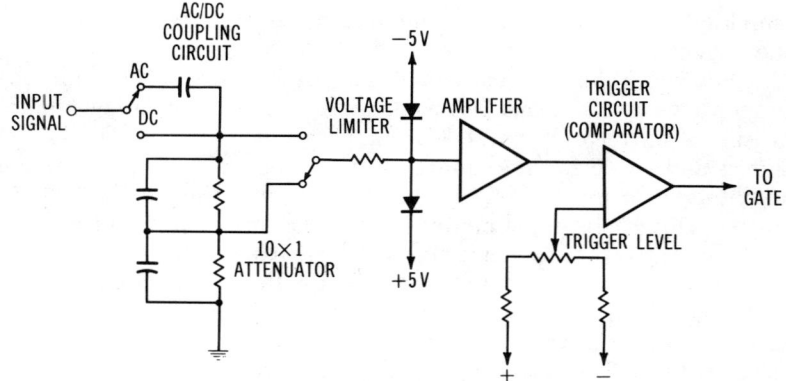

Fig. 2-3. A counter input circuit.

The input attenuator is typically nothing more than a resistive voltage divider that has been properly frequency compensated with shunt capacitors. The attenuator permits the counter to deal with a wide range of signal amplitudes. Inputs with very high voltage levels can be readily accommodated by attenuating them to a lower level more compatible with the input circuitry. For low-level inputs, the attenuator can be switched out completely. Again, a front-panel switch allows the user to select several levels of attenuation. A ten to one (10 × 1) attenuation range is typical.

A voltage limiter circuit is usually a part of the input coupling and attenuator circuit. This usually consists of diodes that protect the input circuits from extremely high voltage levels. The voltage limiter also minimizes overloading. Even though the counter may be capable of withstanding high voltage levels, overloading the amplifier or trigger circuits occasionally will cause errors that can affect the accuracy of measurement. In Fig. 2-3, the resistor and two diodes clamp excessively high input voltages to ±5 volts.

The input amplifier is a wide-range amplifier that is used to raise the amplitude of low-level inputs. The gain of this amplifier determines the sensitivity of the counter. It usually determines the lowest-amplitude signals that can be measured by the counter. Usually the input amplifier allows signals as small as one millivolt to be measured. In the simplest counters, the gain of the amplifier is fixed. In more sophisticated counters, variable gain control may be provided. This allows the sensitivity of the counter to be adjusted for optimum measuring results. In some advanced designs, an automatic gain control (agc) feature is added to allow the gain of the amplifier to be adjusted automatically by the level of the input signal. Regardless of its exact configuration, the amplifier is used to increase the input signal level to the point at which the trigger circuit can operate reliably.

The purpose of the trigger circuit is to shape the input signal into the proper logic levels so that it is compatible with the remaining counter circuits. The trigger circuit takes sine waves or other nondigital input signals and converts them into high-speed logic pulses of the proper amplitude. The most widely used trigger circuit is the popular Schmitt trigger. This can be implemented in several forms. In its simplest form, the Schmitt trigger is a standard digital logic IC. This Schmitt trigger is essentially a bistable multivibrator circuit with feedback that provides very high-speed switching at specific input voltage levels. In its most sophisticated form, the Schmitt trigger may be a high-speed comparator circuit with feedback that can convert input signals of virtually any shape into high-speed logic pulses. If a comparator is used, a variable voltage is usually applied to one input, thereby allowing the user to adjust the trigger level, which determines the point on the input signal at which triggering occurs.

Fig. 2-4 shows an example of how the input circuit processes and conditions an input signal.

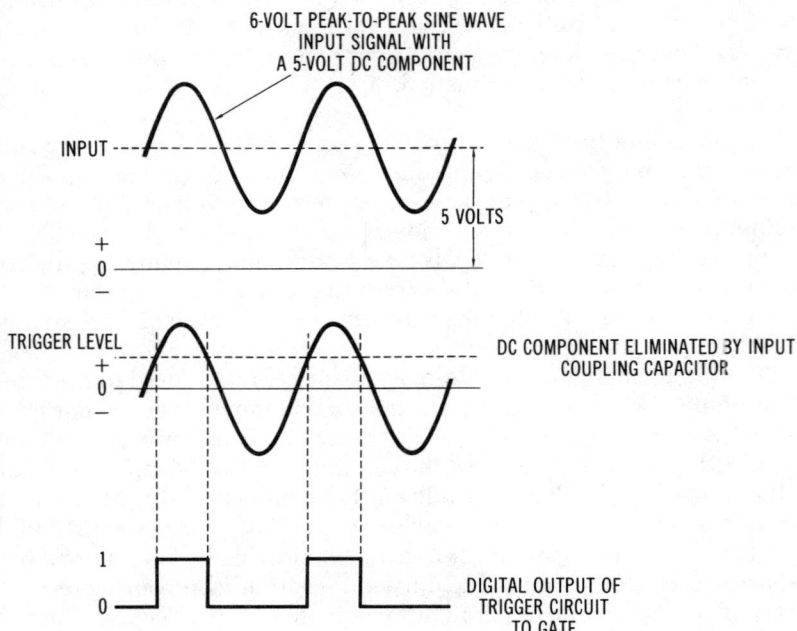

Fig. 2-4. How the input circuit processes a signal to be measured.

## The Main Gate

The trigger output of the input circuit is applied to the main gate. This is a logical AND gate that controls the passage of the input signal to the counter section. Depending upon the mode of operation of the counter, the input signal either will be counted by the main counter section or will be used to control the counting of highly accurate timing pulses.

When the main gate is enabled, or turned on, the conditioned input signal is passed through to the counter. When the gate is disabled, the input pulses are inhibited. The control circuits determine how the gate will be used and determine when it will be enabled or disabled. Fig. 2-5 shows a typical gate circuit with input and output waveforms.

## The Counter Section

The counter section gives the instrument its name. It contains the circuits that count or accumulate the input pulses passed through the main gate. This section of the counter typically consists of a cascaded chain of bcd or decade counters. Each decade is capable of counting by ten and providing bcd outputs of the decimal numbers 0 through 9. Fig. 2-6 shows a typical decade counter section. Depending upon the size, complexity, and cost of the counter, anywhere from four to ten bcd counter circuits are cascaded. The number of counter decades, in turn, determines the number of decimal digits displayed.

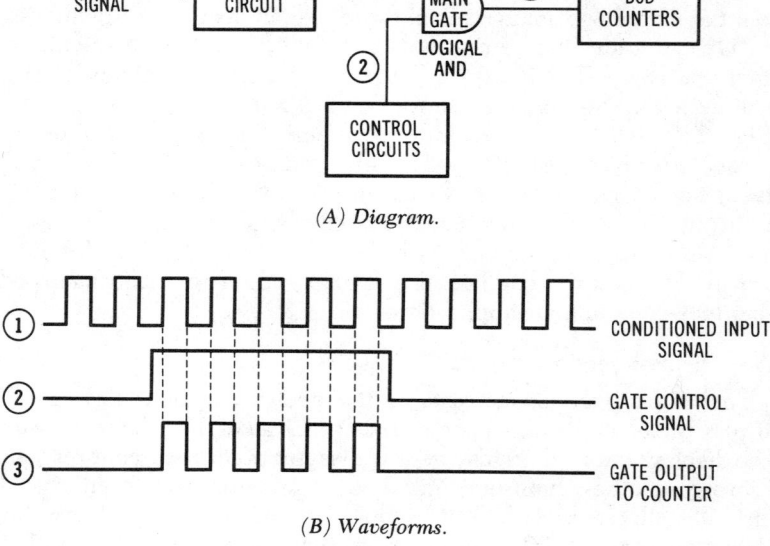

Fig. 2-5. Counter gate circuit.

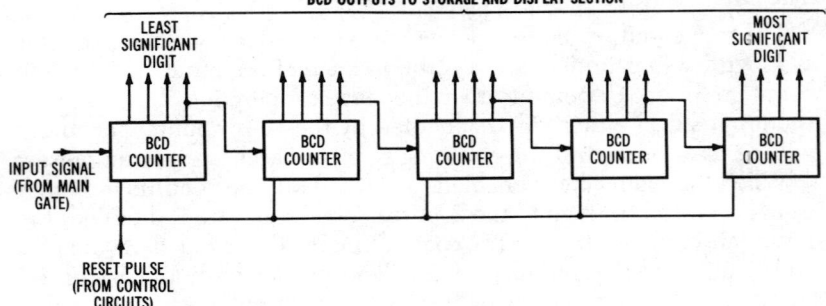

Fig. 2-6. A five-decade bcd counter.

As with the main gate circuit, the decade counters are under the command of the control circuits. The control circuits determine when the main gate will be opened to apply pulses to the counter section. The control circuits also supply a reset pulse so that all decade counters can be reset (set to zero) prior to counting the input.

In some special counters, the bcd decades can also be preset. That is, a specific bcd code can be stored in the counter before the input signal is counted. In other counters, each decade can count either up or down. Most bcd counters count up; that is, as each input pulse is applied, they are incremented so that they count 0, 1, 2, 3, etc. A bcd down counter, on the other hand, is decremented as each input signal is applied. Decade down counters count in the sequence 9, 8, 7, 6, 5, etc. The down count ability allows many special timing functions to be implemented.

Many decade counters used in modern test instruments are single-package MSI integrated circuits. Each IC contains a single complete bcd counter. An example is the 7490 series TTL bcd counter (Fig. 2-7). This decade counter contains four flip-flops internally connected to form a divide-by-2 counter and a divide-by-5 counter. For use as a bcd decade counter, the BD input is externally connected to the A output. The input signal to be counted is applied to input A, and the count sequence is as shown in the bcd truth table in Fig. 2-7B. Inputs are provided to allow resetting all outputs to logical zero or resetting to bcd 9 for nines-complement applications.

## The Display Section

The display section consists of the decimal readout devices that display the contents of the bcd counters plus all of the related circuitry. The displays themselves may be seven-segment incandescent readouts, fluorescent tubes, light-emitting diodes, gas tubes, or liquid crystals. Virtually all types of display readouts are used in digital counters. However, because of their low cost and versatility, seven-segment LED displays are the most popular.

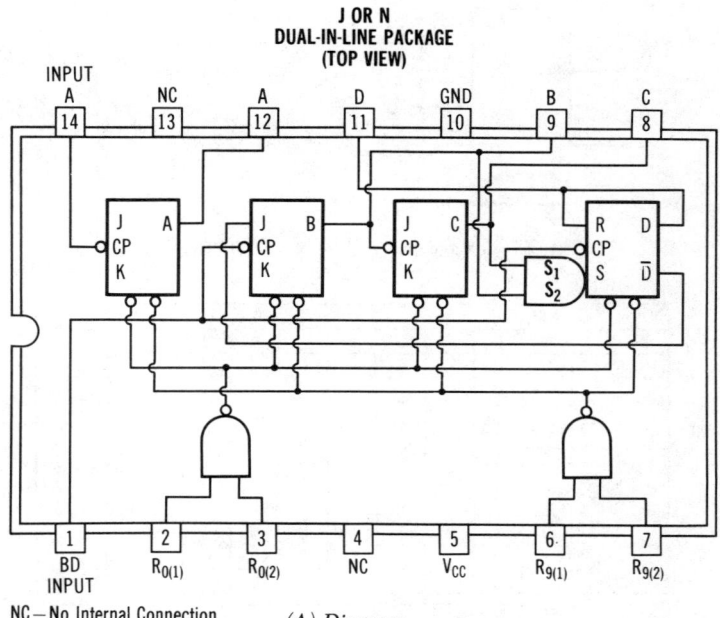

(A) Diagram.

NC — No Internal Connection

### BCD COUNT SEQUENCE

| COUNT | OUTPUT | | | |
|---|---|---|---|---|
| | D | C | B | A |
| 0 | 0 | 0 | 0 | 0 |
| 1 | 0 | 0 | 0 | 1 |
| 2 | 0 | 0 | 1 | 0 |
| 3 | 0 | 0 | 1 | 1 |
| 4 | 0 | 1 | 0 | 0 |
| 5 | 0 | 1 | 0 | 1 |
| 6 | 0 | 1 | 1 | 0 |
| 7 | 0 | 1 | 1 | 1 |
| 8 | 1 | 0 | 0 | 0 |
| 9 | 1 | 0 | 0 | 1 |

### RESET/COUNT

| RESET INPUTS | | | | OUTPUT | | | |
|---|---|---|---|---|---|---|---|
| $R_{0(1)}$ | $R_{0(2)}$ | $R_{9(1)}$ | $R_{9(2)}$ | D | C | B | A |
| 1 | 1 | 0 | × | 0 | 0 | 0 | 0 |
| 1 | 1 | × | 0 | 0 | 0 | 0 | 0 |
| × | × | 1 | 1 | 1 | 0 | 0 | 1 |
| × | 0 | × | 0 | COUNT | | | |
| 0 | × | 0 | × | COUNT | | | |
| 0 | × | × | 0 | COUNT | | | |
| × | 0 | 0 | × | COUNT | | | |

× indicates that either a logical 1 or a logical 0 may be present.

(B) Truth tables.

Courtesy Texas Instruments Inc.

Fig. 2-7. Type 7490 TTL MSI bcd counter.

The display device is typically driven by a decoder/driver circuit. This circuit, usually a single IC, contains logic circuitry that converts the bcd input code into the appropriate logic signals, usually the seven-segment code. Then, driver circuits operated by the decoder output logic signals are used to control the elements of the display. See Fig. 2-8.

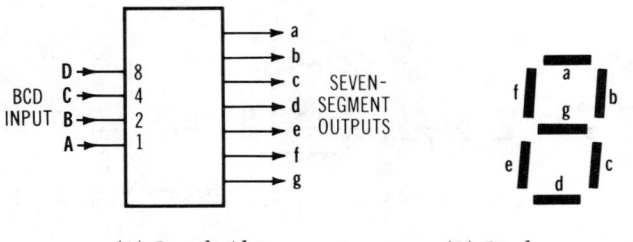

(A) Decoder/driver.      (B) Display segments.

| INPUTS | | | | | SEGMENT OUTPUTS | | | | | | |
|---|---|---|---|---|---|---|---|---|---|---|---|
| DECIMAL | A | B | C | D | a | b | c | d | e | f | g |
| 0 | 0 | 0 | 0 | 0 | 0 | 0 | 0 | 0 | 0 | 0 | 1 |
| 1 | 0 | 0 | 0 | 1 | 1 | 0 | 0 | 1 | 1 | 1 | 1 |
| 2 | 0 | 0 | 1 | 0 | 0 | 0 | 1 | 0 | 0 | 1 | 0 |
| 3 | 0 | 0 | 1 | 1 | 0 | 0 | 0 | 0 | 1 | 1 | 0 |
| 4 | 0 | 1 | 0 | 0 | 1 | 0 | 0 | 1 | 1 | 0 | 0 |
| 5 | 0 | 1 | 0 | 1 | 0 | 1 | 0 | 0 | 1 | 0 | 0 |
| 6 | 0 | 1 | 1 | 0 | 1 | 1 | 0 | 0 | 0 | 0 | 0 |
| 7 | 0 | 1 | 1 | 1 | 0 | 0 | 0 | 1 | 1 | 1 | 1 |
| 8 | 1 | 0 | 0 | 0 | 0 | 0 | 0 | 0 | 0 | 0 | 0 |
| 9 | 1 | 0 | 0 | 1 | 0 | 0 | 0 | 1 | 1 | 0 | 0 |

BCD      0 = Segment On
     1 = Segment Off

(C) Truth table.

Fig. 2-8. Bcd to seven-segment decoding.

The bcd input to the decoder/driver circuit usually comes from a four-bit storage register that is provided for each bcd counter (Fig. 2-9). The outputs of the bcd counter are connected to the four-bit storage register. When the bcd counter stops counting, its contents are transferred by a logic signal into the four-bit storage register. At this time, the counter may then begin counting again without fear of losing the previous value. The bcd output of the storage register is then applied to the decoder/driver. Therefore, it is the contents of the storage register that are displayed rather than the actual contents of the bcd counter.

The main purpose of the storage register is to eliminate flickering of the display as the counter section is accumulating the input pulses. During the count time, the display digits will be changing very quickly so as to present a continuous and annoying display blur. Of course, it is not possible to read the display during counting. With the storage register, however, this blur and flicker are completely eliminated, and the display always shows a fixed readout.

In most counters, the arrangement shown in Fig. 2-9 is used. Here

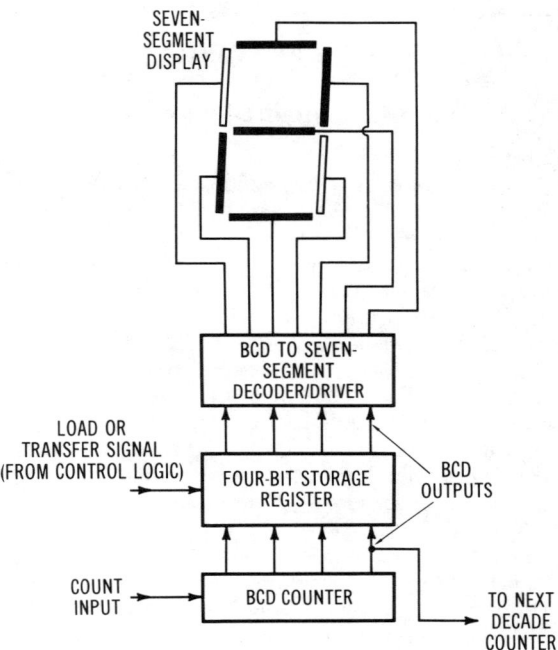

Fig. 2-9. Storage register, decoder/driver, and readout display used with bcd counter.

each decade counter is provided with its own storage register and a separate decoder/driver for each display digit. However, in some counters a special *multiplexed* display arrangement is used, and only a single decoder/driver circuit is needed. Its outputs are applied to all display readout elements simultaneously. A special scanner circuit then sequentially enables each of the readouts in the display one at a time. At the same time, the bcd outputs of the decade counters are sequentially connected, one at a time, to the decoder/driver. The result is the same as the standard display arrangement. However, for counters with many digits, this multiplexed display arrangement uses less circuitry and is much lower in cost. See Reference IV for more details.

## The Control Circuits

The control circuits are really the brains of the counter because they determine how the counter is being used. They are digital logic elements that generate control pulses to operate the main gate, decade-counter, and display sections. In conjunction with the time base, the control circuits generate the gating signal for the main gate. In addition, the control circuits generate the reset pulses for the decade counters and the transfer pulses that cause the contents of the decade

# REFERENCE IV

## How Multiplexed Displays Work

When a counter (or other digital instrument) uses six or more readout displays, it becomes expensive to provide each counter decade with its own decoder/driver circuit. An alternative to this arrangement is the dynamic, or multiplexed, display. In a multiplexed display, each bcd decade in the counter section has its own four-bit storage register and separate display readout. However, only one bcd to seven-segment decoder/driver circuit is used. The outputs of the decoder/driver are connected to all of the segments of every decimal display. An example of this is illustrated in Fig. IV-1. The decoder/driver is used to drive each of the eight LED displays, but not simultaneously. Instead, each seven-segment display is enabled one at a time in sequence by turning on the appropriate driver transistor, $Q_1$, $Q_2$, . . . $Q_8$. This sequencing operation takes place at a high rate of speed and is accomplished by the scan counter and decoder circuit. The scan oscillator generates a clock signal at several kilohertz. Alternatively, the clock signal can be taken from the time-base circuit in the counter. The clock increments a three-bit binary counter. The counter outputs are decoded into eight signals by the one-of-eight, or octal, decoder circuit. Only one of the eight output lines is on at a time; thus each display is turned on by itself for a short period of time. Since the display is on for only a period of microseconds, ordinarily it would not be seen. But each display is turned on repeatedly every eight counts of the scan clock. For example, if the scan oscillator frequency is 8 kHz, each display will be enabled for 125 microseconds every 1000 microseconds. The display will be dimmer with this arrangement than it would be if it were on continuously; however, normal display brightness is obtained by compensating with higher currents through the display elements. Since the displays are being turned on and off so rapidly, the normal human eye cannot follow, and the appearance is one of a continuous display. The persistence of vision of the eye permits this multiplexed arrangement to be practical.

The contents of the bcd counters are to be displayed. The numbers in the counters are first transferred to the four-bit storage registers. The register outputs are then applied to sets of multiplexer gates. These sets of four gates permit the bcd outputs of each storage register to be applied to the bcd to seven-segment decoder/driver, one at a time. The outputs of all eight sets of multiplexer gates are tied together to create a wired OR. Three-state output gates can be used in this application. Each set of multiplexer gates is enabled one at a time by the scan decoder outputs. For example, to display the contents of the first (No. 1) bcd counter, its contents are loaded into the first four-bit storage register. When scan decoder output No. 1 is a logical 1, it enables the No. 1 multiplexer gates, thereby applying the bcd contents of the storage register to the

*Continued on page 32.*

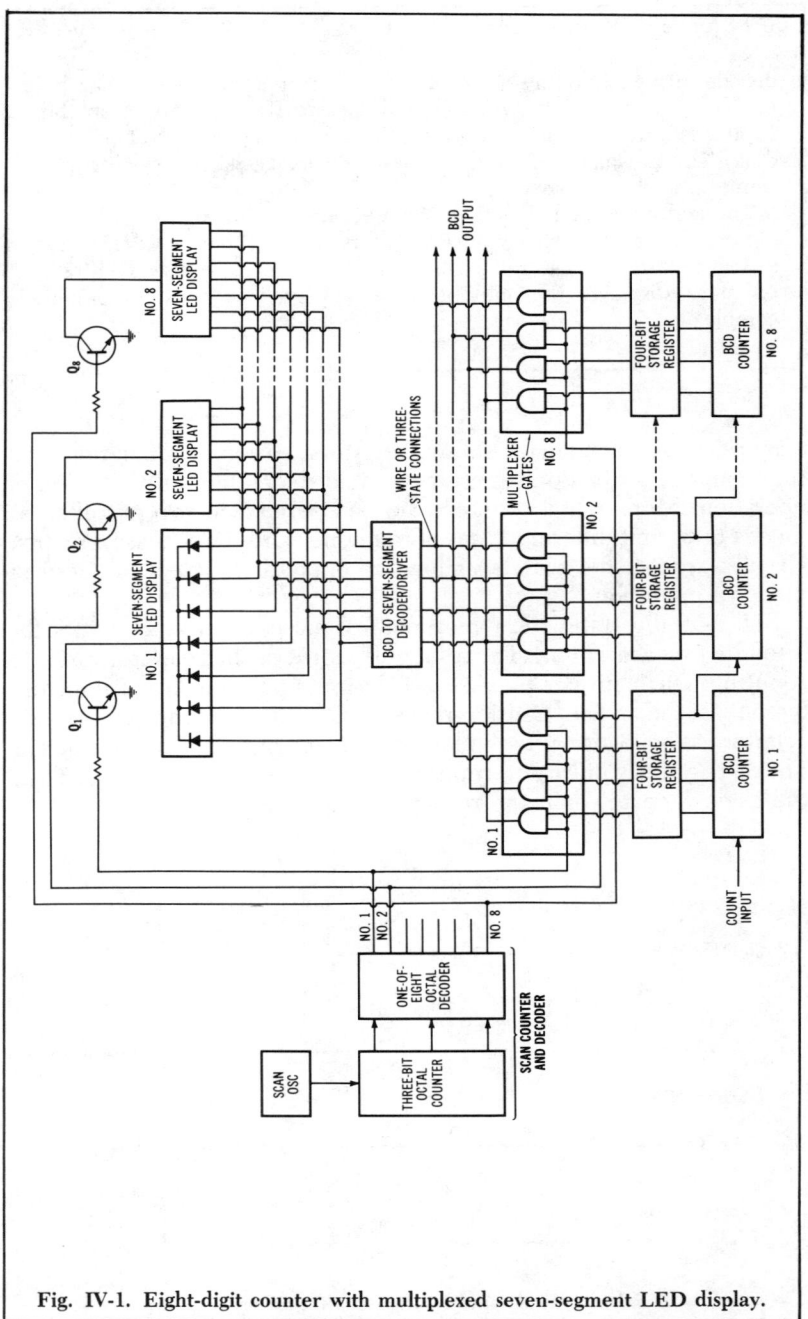

Fig. IV-1. Eight-digit counter with multiplexed seven-segment LED display.

> REFERENCE IV (cont.)
>
> decoder/driver. Also, the No. 1 scan decoder output turns on $Q_1$, thereby enabling LED readout No. 1. The decoder/driver then turns on the appropriate segments in the readout, thus displaying the correct decimal digit. This action is repeated at a fast rate for each of the digits and counters in the display.
>
> The multiplex display method is not economical for counters with six digits or fewer. Therefore, in small counters each bcd counter will have its own decoder/driver and related circuitry. But in larger counters with six or more displays, the multiplex method reduces cost, size, and complexity.

counters to be loaded in the storage registers. Various combinations and sequences of control pulses are generated depending on the mode of operation. Most digital counters can be operated in several different modes to permit different types of frequency and time measurements. Usually, front-panel switches allow the operator to select the desired mode of operation.

Fig. 2-10 illustrates the typical control pulses used in a counter for frequency measurement. The reset pulse zeros the bcd counters prior to counting. The main gate pulse allows the input pulses to be accumulated for one second. The transfer pulse causes the contents of the bcd counters to be loaded into the four-bit storage registers and displayed. The display delay pulse determines the length of the display time. Then the cycle repeats when the reset pulse again occurs.

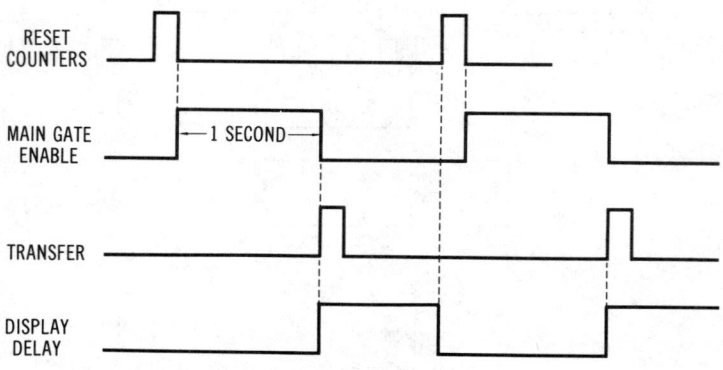

Fig. 2-10. Typical control pulses in a frequency counter.

## The Time Base

The time base is a highly stable oscillator circuit that is used to generate the timing pulses used by the counter. These timing pulses must be extremely precise if high-accuracy time and frequency measurements are to be made. It is the time base that determines the overall accuracy of the counter.

The time base in most digital counters is a precision crystal oscillator. This oscillator usually generates a signal somewhere between 1 MHz and 10 MHz. This high-frequency signal is then divided by bcd counters used as decade ($\div 10$) frequency-divider circuits as shown in Fig. 2-11. The time base, therefore, delivers a wide range of fixed-frequency, high-accuracy signals. These are applied to the control circuits and may be used in a variety of ways. On many counters, a front-panel control allows the user to select the time-base frequency to be used. This, in turn, controls the resolution of the time or frequency measurement.

The quality of the crystal itself and its temperature stability determine the accuracy of the counter. The frequency of the crystal must remain stable if reliable and accurate measurements are to be made. However, most crystals are extremely temperature sensitive, and even minor variations in temperature affect their frequency. For simple low-cost counters to be used in noncritical applications, an ordinary crystal may be used. However, most good bench test instruments use temperature-compensated crystal oscillators (tcxo) to ensure that the frequency remains stable over a wide range of temperature variations. A further improvement in accuracy and stability can be obtained by using a counter with a crystal housed in a temperature-controlled oven. Here, the temperature of the crystal can be maintained within several tenths of a degree. In effect, the crystal temperature is essentially fixed, thereby providing a constant frequency output from the time base. Such oven-controlled crystals are used in the best of laboratory counter instruments.

For even greater accuracy and stability, higher-precision time bases are available. An example is the cesium-beam atomic standard available in some very high cost counters.

For noncritical applications, it may not even be necessary to use a crystal-controlled time base. In some applications, a sufficiently accurate measurement can be obtained simply by using the ac power line as a frequency standard. The frequency of most ac power lines is controlled to better than 0.1%. It is easy to use the 60-Hz power-line signal and convert it into time-base signals of 100 milliseconds, 1 second, and 10 seconds with appropriate frequency dividers.

## Integrated-Circuit Counters

In the past, most counters were constructed from conventional SSI and MSI integrated circuits. These TTL, CMOS, and in some cases

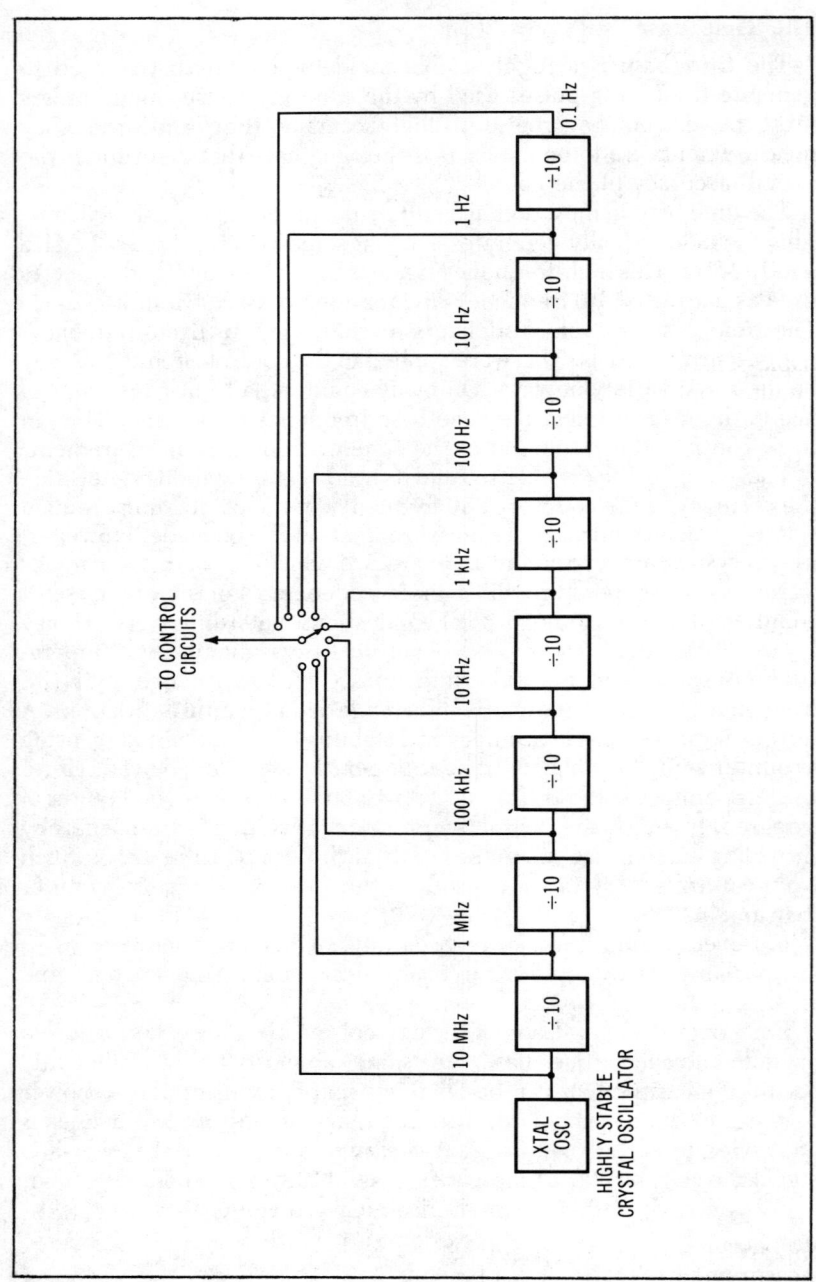

Fig. 2-11. Time-base circuitry.

ECL, devices were combined to form all of the previously described circuits. But today, thanks to advanced semiconductor technology, virtually all counter circuits can be made as a single unit on one chip of silicon. Large-scale integration (LSI) of MOS circuits now permits an entire counter—with the exception of the crystal, the display elements, the various switches or controls, and the power supply—to be contained in one 28- or 40-pin dual in-line package.

Fig. 2-12 is a block diagram of a typical LSI counter. This unit contains eight bcd counters with their related storage registers (latches) and multiplexer gates; the scan oscillator, counter, and decoder for the multiplexed display; the seven-segment decoder/driver; and all of the display enable (strobe) circuits. This counter does not contain the input circuit, control circuits, or time base, which can be separately customized to the specific counter. This LSI device has a typical 5 MHz upper counting frequency.

A complete counter on a chip, including the time base, input circuit, and control circuits, is illustrated in Fig. 2-13. The counter has an eight-digit capability with a multiplexed display. An external 10-MHz crystal is needed for the time base. The upper frequency counting limit is 10 MHz. With devices of this kind, counters made with separate MSI ans SSI ICs are a thing of the past.

## COUNTER OPERATIONAL MODES

The mode of a digital counter determines what function the counter performs. Most general-purpose digital counters have several modes so that various types of frequency and timing measurements can be made. Usually a front-panel switch is used to select the operating mode.

### Frequency-Measurement Mode

The most commonly used counter mode is frequency measurement. In fact, most digital counters are usually referred to as frequency counters. Many low-cost and special counters have only a frequency-measurement mode. Frequency is such a common and important characteristic of electronic signals that it is impossible to design, operate, or service most electronic equipment adequately without a frequency measuring capability. The need is particularly critical in the radio-frequency spectrum. Modern counters are capable of measuring frequencies from subaudio to microwave. Table 2-1 shows the various frequency bands.

Frequency is a measure of the number of events or cycles of a signal that occur in a given period of time. The usual unit of frequency measurement is the cycle per second, or *hertz* (Hz). The way a counter measures frequency is to count the number of cycles of the input signal that occur in a known time period. The time base generates a very

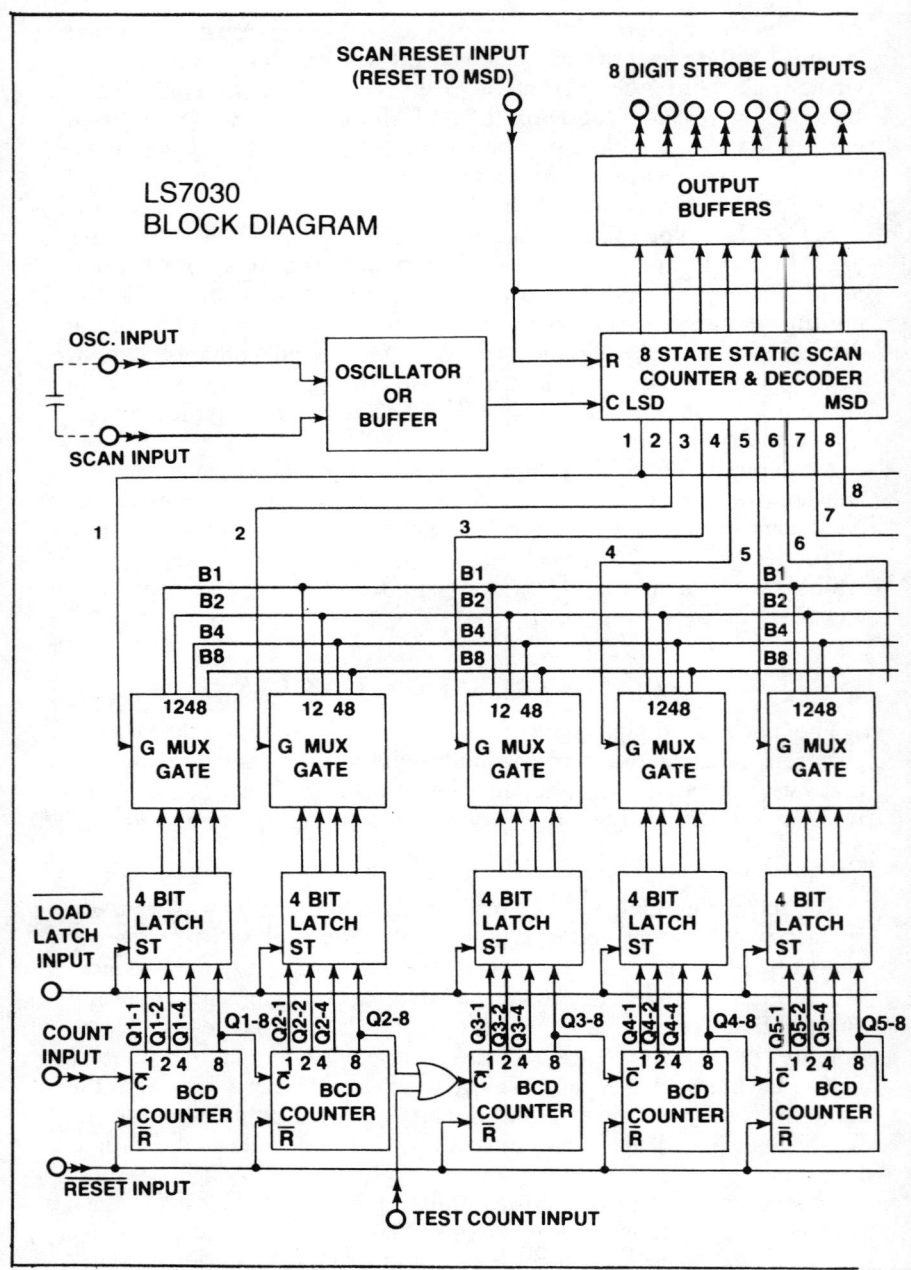

Fig. 2-12. A p-channel

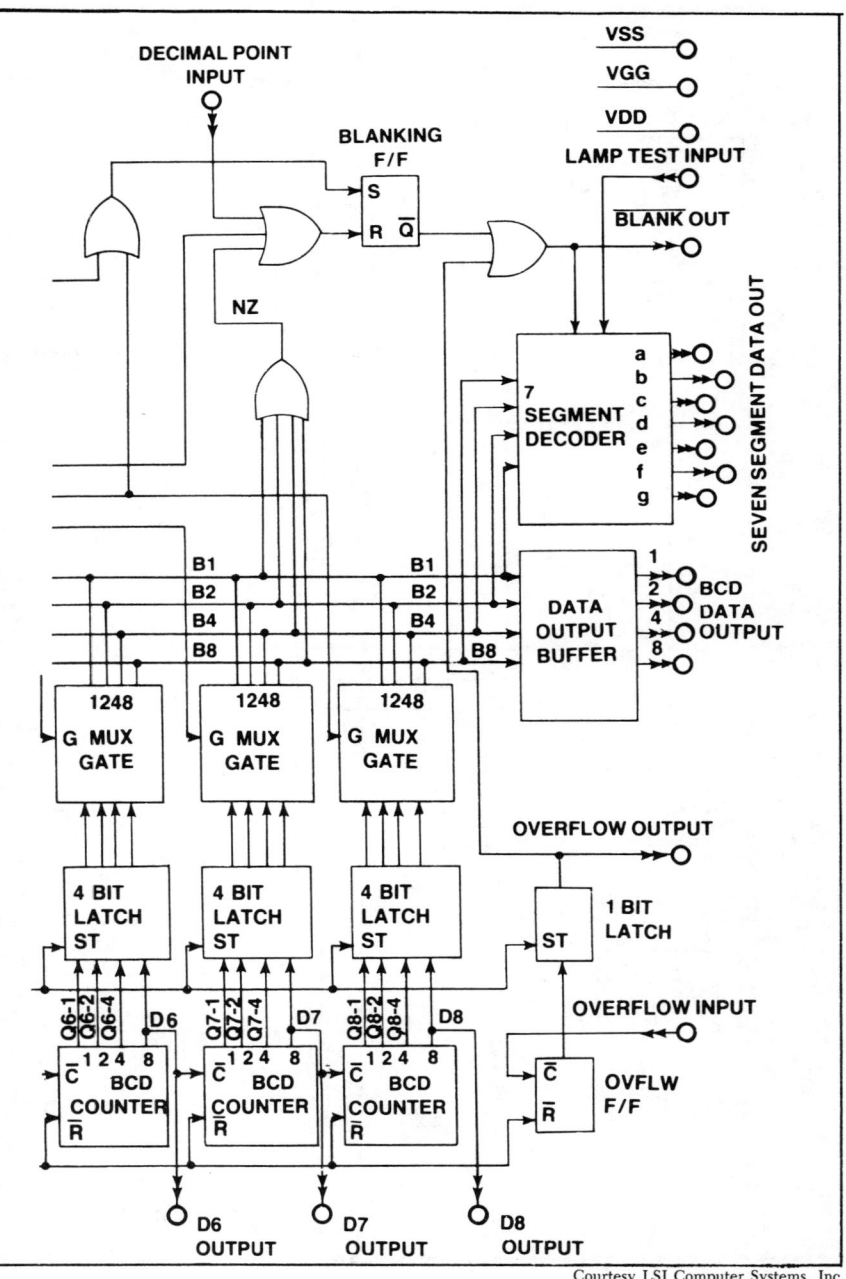

MOS LSI counter.

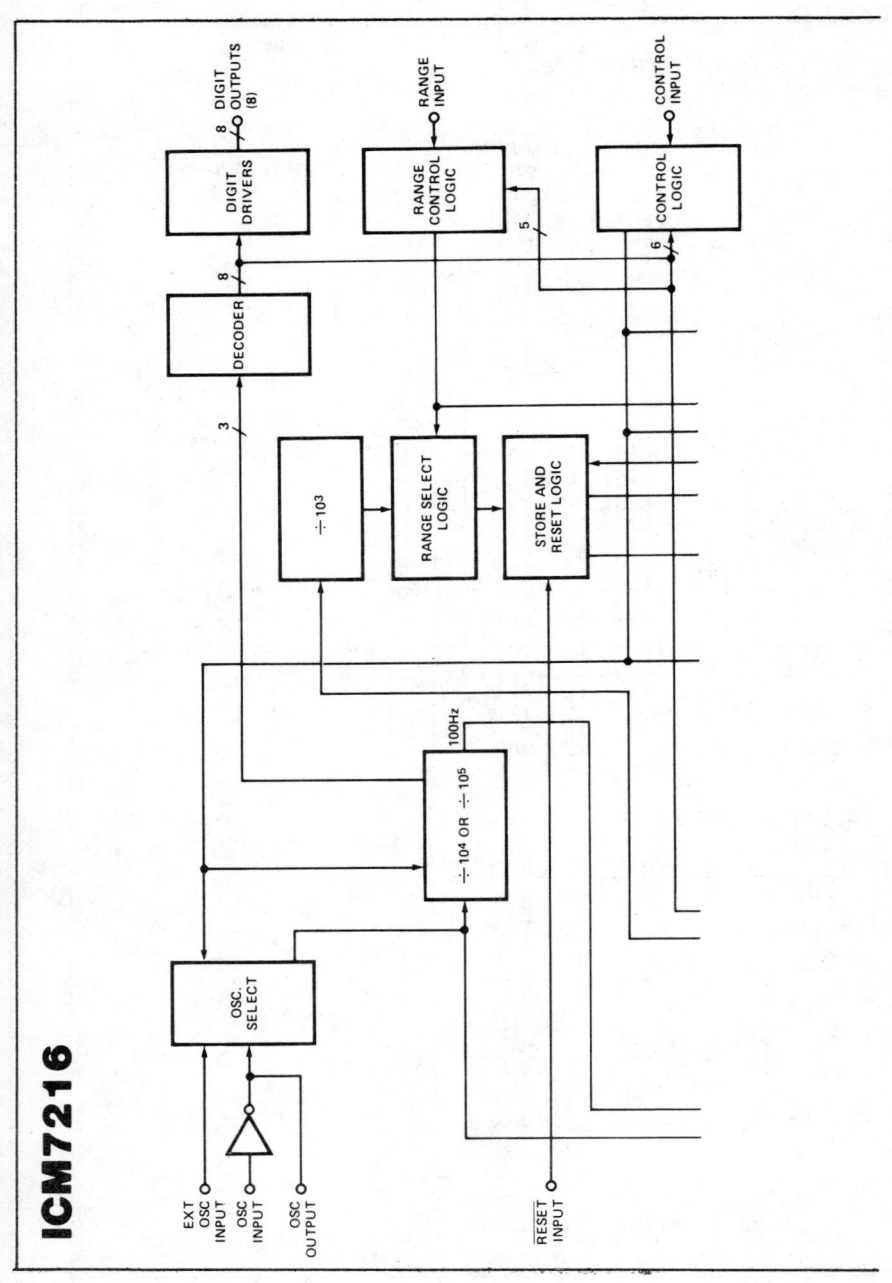

Fig. 2-13. A CMOS LSI counter for

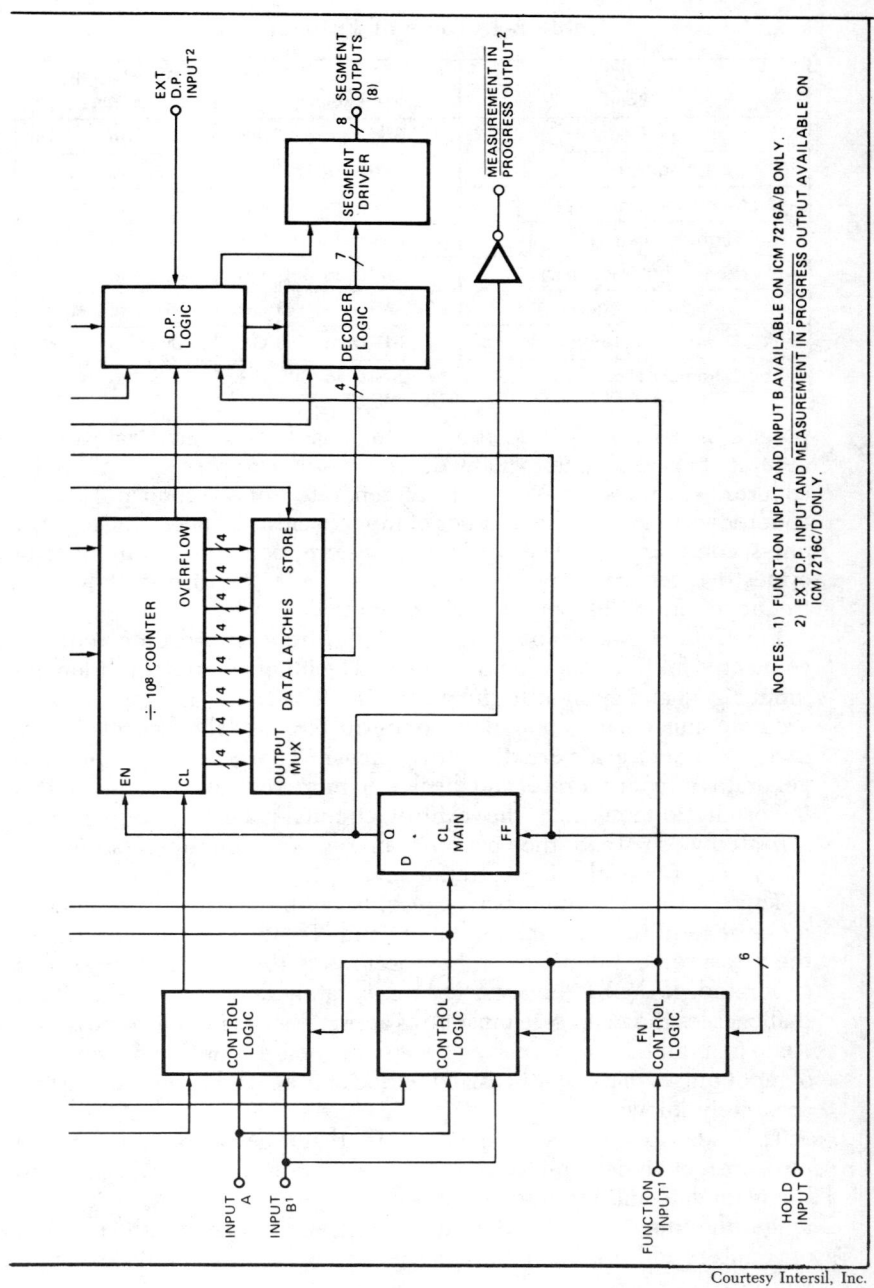

time and frequency measurements.

Table 2-1. Table of Frequencies

| Band | Frequency | Wavelength (λ) in Meters |
|---|---|---|
| vlf (Very Low Frequency) | 3 kHz to 30 kHz | 100 km to 10 km |
| lf (Low Frequency) | 30 kHz to 300 kHz | 10 km to 1 km |
| mf (Medium Frequency) | 300 kHz to 3 MHz | 1 km to 100 m |
| hf (High Frequency) | 3 MHz to 30 MHz | 100 m to 10 m |
| vhf (Very High Frequency) | 30 MHz to 300 MHz | 10 m to 1 m |
| uhf (Ultra High Frequency) | 300 MHz to 3 GHz | 1 m to 10 cm |
| shf (Super High Frequency) | 3 GHz to 30 GHz | 10 cm to 1 cm |
| ehf (Extremely High Frequency) | 30 GHz to 300 GHz | 1 cm to 10 mm |

precise signal that is used to open, or enable, the main gate for an accurate period of time to allow the input pulses to pass through to the counter. If this is a very precisely generated one-second pulse, the counter accumulates the number of input cycles that occur during that one-second interval. The counter, therefore, contains the number of cycles that occurred in one second. As a result, the display shows the frequency in cycles per second, or hertz.

Fig. 2-14 shows a block diagram of the frequency-counter circuits connected for frequency measurement. The input circuits condition the input signal and apply it to the main gate. The time base generates an accurate signal that is applied to the control circuits. The control circuits in turn generate a periodic gating signal. The control circuits also generate the counter reset and display memory register load pulses. It is important to note that the control circuits generate these signals repeatedly so that the counter makes a continuous series of measurements of the input frequency.

There are two important concepts with regard to frequency measurement that you should understand. The first is that the truth of the measurement depends on how accurately the gate time interval is generated. If the frequency is to be accurately determined, the signal that enables the main gate must be as accurate and as stable as possible. Keep in mind that in measuring frequency we are counting the number of input pulses that occur in a fixed period of time, and that time must be accurately known.

The second important concept is that the resolution of the measurement is determined by the time duration of the signal applied to the main gate and the number of bcd counters and display digits. The longer the time interval of the gating pulse and the greater the number of counters and display digits, the better the measurement resolution will be.

To illustrate the concept of resolution, let us assume that we are going

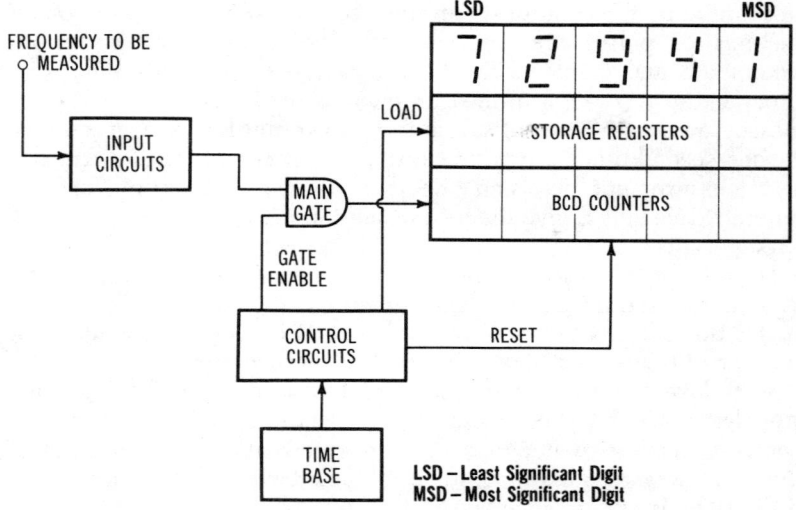

Fig. 2-14. Counter circuits connected for frequency measurement.

to measure the frequency of an input signal whose frequency is 65,536 Hz, or 65.536 kHz. We will also assume that the counter has five bcd decades and five digits of display.

Now assume that we select a 1-Hz time-base output signal. The period (t) of the 1-Hz signal, of course, is one second (t = 1/f). This 1-Hz signal is applied to the control circuits, which in turn generate a precise one-second gating signal for the main gate. This will allow the input signal to be applied to the counter for one second. The measurement cycle goes something like this: The control circuit first generates a reset pulse to set all the decade counters to zero. Next, the one-second gating signal enables the gate, allowing the input pulses to be counted by the decade counters. After the one-second gating interval, the main gate is inhibited. The control circuit then generates a pulse that causes the contents of the bcd counters to be transferred to the display storage registers. The display then shows the frequency of the measured signal. In this case, since the gate was open for one second, the counter will have accumulated 65,536 input pulses. The five-digit display indicates the measured frequency.

Assume now that a 100-Hz time-base signal is selected instead of the 1-Hz signal. This means that the control circuit will generate a 10-millisecond gating pulse instead of the one-second pulse (t = 1/f = 1/100 = 0.01 second = 10 milliseconds). In other words, the input signal will be allowed to be applied to the decade counters only for the precise interval of ten milliseconds. Since 10 milliseconds is one

one-hundredth of a second, only one-hundredth of the input pulses will be counted by the counter. One-hundredth of 65,536 is 655.36. Since fractions of pulses are not counted, the counter would simply accumulate and display 655. As you can see, with this time-base interval, the resolution of measurement is to the nearest 100 hertz. Inexact measurement results from the poor resolution. As you can see, the one-second time base produces the greater resolution and thus the best measurement precision. Keep in mind that the time base is generally selected to give the best resolution with the given number of display digits.

The number of decade counters and display digits also determines the resolution of the frequency measurement. The greater the number of digits, the better the resolution will be. However, the greater the number of digits, the higher the cost of the counter will be.

Most low-cost counters have at least five digits of display. This provides reasonably good resolution on most frequency measurements. For very high frequency measurements, resolution is more limited. However, good resolution can still be obtained with a minimum number of digits by optimum selection of the time-base frequency.

Suppose that the time base of a five-digit counter is set to 1 Hz, thereby creating a one-second gate pulse. Assume that a frequency of 4,903,156 Hz is applied to the input. During the one-second gating period, 4,903,156 input pulses will occur. These, of course, will be counted by the counter. Keep in mind that there are only five decade counters and display digits, and the maximum count capability is 99,999. As the input pulses occur, during the one-second counting period, the fifth, or most significant, counter will overflow many times. Nevertheless, the five counters will still contain the correct number of counted pulses. In this case, the most significant digits of the input frequency, 4 and 9, will not be displayed. However, the last five digits of the input frequency will be displayed correctly. The display will read 03156. The resolution is 1 Hz because of the one-second gating pulse. Seven digits are required to display the full frequency with a resolution of 1 Hz.

Now, if a shorter time base is selected, for example 1 millisecond, the frequency will be measured with less resolution, but the two most significant digits of the frequency will be displayed. Selecting a 1-kHz time-base signal will create a 1 millisecond gate pulse. During this 1-millisecond gate period, 4903 input pulses will occur. The display will read 4903. This gives you the first two digits of the frequency, but the resolution is only 1 kHz.

Reference V further elaborates on the subject of resolution and precision. Table 2-2 shows the relationship between the time-base frequency and resolution of measurement.

In most counters that have a selectable time base, the position of the

# REFERENCE V

## Precision, Resolution, And Significant Figures

Precision is the quality of being sharply or exactly defined. It can be thought of as the resolution with which a quantity is measured. Resolution is the measure of the smallest increment in change of a quantity that can be distinguished. A measured frequency of 341 kHz could be thought of as having a resolution to the nearest 1 kHz. Our counter may not be capable of resolving any smaller frequency increment and, therefore, cannot provide any further precision. A better counter may reveal that this same frequency is 341.275 kHz, or 341,275 Hz. We now have more significant figures and, therefore, greater resolution and precision. The frequency is more precisely defined because the counter can resolve to the nearest hertz. Greater resolution results in more significant figures and, therefore, greater precision.

Significant figures are those digits in a number that actually convey information about the magnitude of the quantity being measured. The number of figures used to represent the quantity also indicates the precision with which the quantity is expressed. The greater the number of significant figures, the greater the precision is.

Do not confuse precision with accuracy. They are not the same. Accuracy means closeness to the truth. Your counter may give you fine resolution and many significant digits, but it may be inaccurate. The time base may be incorrect so that, despite the resolution, the frequency the counter indicates is something considerably different from the true frequency.

Table 2-2. Relationship Among Time-Base Frequency, Gate Time, and Measurement Resolution

| Time-Base Frequency (f) | Gate-Pulse Time (t = 1/f) | Measurement Resolution | Position of Display Decimal Point | Display Reads |
|---|---|---|---|---|
| 0.1 Hz | 10 seconds | 0.1 Hz | 0000000.0 | Hertz |
| 1 Hz | 1 second | 1 Hz | 00000000. | Hertz |
| 10 Hz | 100 milliseconds | 10 Hz | 000000.00 | Kilohertz |
| 100 Hz | 10 milliseconds | 100 Hz | 0000000.0 | Kilohertz |
| 1 kHz | 1 millisecond | 1 kHz | 00000000. | Kilohertz |
| 10 kHz | 100 microseconds | 10 kHz | 000000.00 | Megahertz |
| 100 kHz | 10 microseconds | 100 kHz | 0000000.0 | Megahertz |
| 1 MHz | 1 microsecond | 1 MHz | 00000000. | Megahertz |

display decimal point is automatically adjusted as the time-base signal is adjusted. In this way, the display always shows the frequency in units of hertz, kilohertz, or megahertz.

In some counters, only two time-base frequencies, 1 Hz and 1 kHz, are used for frequency measurement. This provides gating pulses of one second and one millisecond. These two gating intervals provide good resolution over a wide frequency range with a minimum of circuit complexity and expense.

Some of the more sophisticated counters have an automatic time-base selection feature called *auto-ranging*. Special auto-ranging circuitry in the counter automatically selects the best time-base frequency for maximum measurement resolution without *over-ranging*. Over-ranging is the condition that occurs when the count capability of the counter is exceeded during the count interval. The number of counters and display digits determines the count capability and thus the over-range point for a given time base.

**Ratio Mode**

The ratio mode is used to compare the frequencies of two signals. In this mode, the internal time base is not used. Instead, one of the signals is applied to an external input and replaces the time-base signal that drives the control circuits. The other signal whose frequency is to be compared is applied to the standard counter input. Otherwise, the circuits are the same as those used in the normal frequency-measuring mode. See Fig. 2-15. When switch SW1 is in the A position, the normal counter time base is connected to the control circuits, and the counter is in the normal frequency-measurement mode. If SW1 is set to the B position, the external time-base input ($f_2$) is connected to the control circuits. Switch SW2 will normally be in the A position (SW2 and the three bcd counters will be explained later). The result of this configuration is that the frequencies of the two input signals are compared, and the counter displays a number ($f_2/f_1$ or $f_1/f_2$) that is the ratio of the two frequencies.

This measurement mode accomplishes the frequency comparison by allowing one signal to control the main gate for another. The signal connected to the external time-base input is used by the control circuits to create a gating pulse that is used to enable the main gate. Typically, this gating signal will have a time duration equal to one period of the signal applied to the external time-base input. When the gate is enabled, cycles of the second input signal are counted by the counter. The number of pulses that occur during the one-period interval of the first input signal is numerically equal to the ratio of the two frequencies. Fig. 2-16 illustrates this. In this example, the period of $f_2$ is seven times that of $f_1$; therefore, $f_1/f_2 = 7/1$. Naturally, with this arrangement, the higher-frequency signal ($f_1$) should be applied to the regular counter

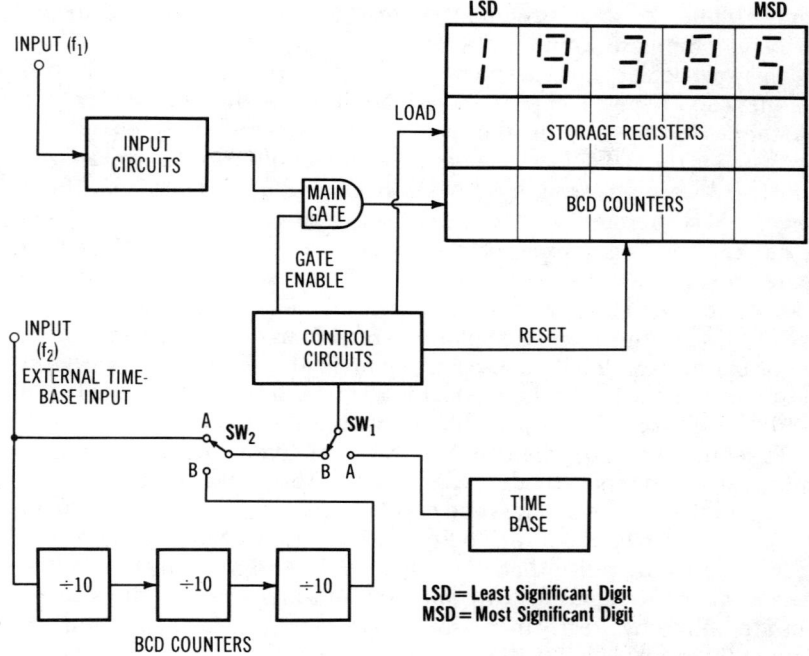

Fig. 2-15. Counter circuits connected for the ratio mode.

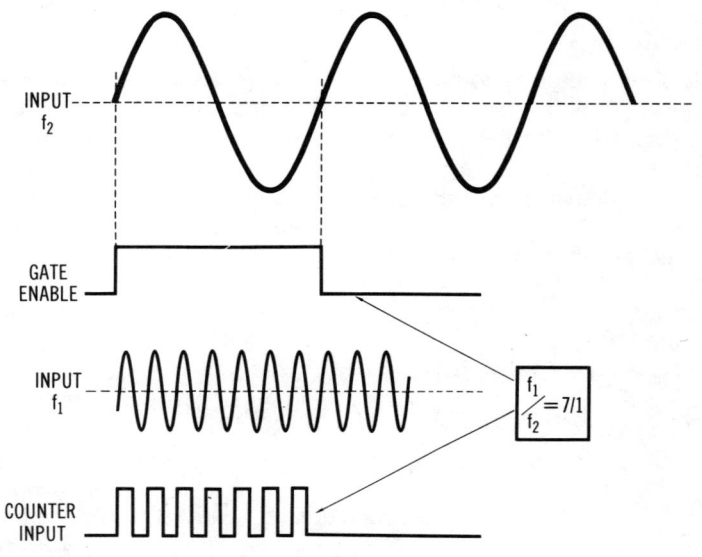

Fig. 2-16. Counter signals in the ratio mode.

input, while the lower-frequency signal ($f_2$) should be applied to the external time-base input.

This particular measurement mode works well when the two frequencies being compared are significantly different from one another. However, when the frequencies become nearly equal, the precision of the ratio measurement deteriorates quickly. As an example, consider the comparison of a 1-MHz signal and a 7-MHz signal. If the lower 1-MHz frequency is applied to the external time-base input, it will generate a gating period of one microsecond. The 7-MHz signal generates pulses that occur every 143 nanoseconds. During the one-microsecond gating interval, seven of the 7-MHz pulses will be counted. Thus the counter display will read 7. In most counters there is a normal plus or minus one-count error in the least significant digit position. (This will be explained later.) This means that the actual ratio could be 6 or 8. This represents a possible error of ±14.3%.

To reduce the error when the signals are relatively close in frequency, some extra frequency dividers can be used. The lower input frequency can be applied to the dividers and divided down by a factor of 10, 100, or 1000. The three bcd counters in Fig. 2-15 are used for this purpose. The result is a gating pulse that is 10, 100, or 1000 times longer than the period of input signal $f_2$. This allows many more input pulses to be counted, thereby greatly increasing the resolution of the measurement. The number displayed is the ratio $Xf_1/f_2$, where X is the division factor (10, 100, or 1000). The display reading can be quickly corrected by dividing by the factor X.

**Period Mode**

It is often necessary to know the period of a signal being measured. The period (t) is the time it takes for the input signal to complete one cycle (Fig. 2-17). Period is the reciprocal of frequency (t = 1/f). If the frequency is known, the period can be quickly computed with an electronic calculator. However, by rearranging the existing circuits in the counter, the period can be very quickly and easily measured and displayed directly. This is convenient in that it eliminates the extra

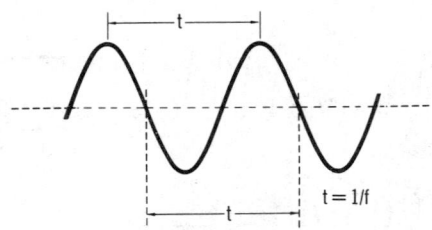

Fig. 2-17. The period (t) of a signal.

time-consuming step of computing the period from the known frequency.

Another advantage of the period mode is that it provides greater accuracy of measurement for low-frequency signals. For example, assume that you wish to measure a low-frequency signal such as the voltage from a 60-Hz power line. With most counters, you would select a 1-second gate period, and the counter would display the number 60. Sixty cycles of the input signal occur during the one-second gating period; thus, the display indicates the frequency to the nearest 1 Hz. The resolution of this measurement then is no better than 1 out of 60, or ±1.67%. This is not a very accurate measurement of power-line frequency.

One way to increase the accuracy of such low-frequency measurements is to increase the gate interval. This can be done by providing one extra decade frequency divider in the time-base divider chain. In this way, a time base output of 0.1 Hz can be obtained. When applied to the control circuitry, the gate-enable pulse interval will be 10 seconds. During this ten-second period, the counter will count the incoming low-frequency signal. In the case of a 60-Hz input, 600 cycles would occur; therefore, the counter would read 600, or 60.0 Hz. The least significant digit indicates tenths of a cycle, or 0.1-Hz resolution. As you can see, increasing the measurement interval provides greater resolution of measurement. In this case, the accuracy of measurement becomes 1/600 = 0.167%. This is a considerable improvement in measurement precision over the same measurement with a one-second gate period.

While using a longer gate period improves the measurement precision, the length of time involved in making the measurement is considerably extended. Ten seconds may not seem like a long time, but for a typical electronic measurement it is extremely long. If one is making such measurements repeatedly, they can become extremely boring and time-consuming. Increasing the gate time to 100 seconds would provide a further improvement in measurement resolution and accuracy. However, 100 seconds is better than a minute and a half, and this is an unacceptably long time for a measurement to take place.

To improve the resolution and reduce the time of low-frequency measurements, the period mode can be used. The period of the input signal is measured much more quickly and inherently provides greater resolution.

Again consider the example of measuring the 60-Hz power-line frequency. The period can be measured in milliseconds by counting 1-kHz pulses from the time base. On a five-digit counter, this would provide a reading of 16.666 milliseconds. This represents a significant improvement in the measurement precision. A further improvement can be obtained by counting microsecond pulses. There are 16,666 microseconds per period in a 60-Hz signal. Depending on the number of decade counters and display digits, significant gains can be made in counter measurement of low-frequency signals by measuring the period.

The only drawback to the measurement of low-frequency signals by the period mode is the necessity to convert the measurement into frequency. Again, this can be easily done with an electronic calculator (f = 1/t). If many measurements must be taken, however, the measurement and conversion process becomes time-consuming and tedious.

Several manufacturers have developed computing counters designed specifically for measuring low-frequency signals. In many applications involving audio and subaudio frequencies, high-accuracy low-frequency measurements are often required. In audio applications involving filters and musical instruments, low-frequency measurements must often be made. In the field of geophysics and in the vibration testing of mechanical equipment, low-frequency measurements are also required. These counters perform most standard counting functions but in addition provide a computing capability. The counter measures the frequency of the low-frequency signal by measuring the period. Once the period is measured, the counter automatically computes the frequency and displays it. Special counter circuitry is used to do this. Special logic circuits can be constructed to perform the operation, although in some modern counters MOS LSI calculator chips or microprocessors are used to perform the period-to-frequency computation.

The period-mode circuitry of a digital counter is illustrated in Fig. 2-18. Here the input signal whose period is to be measured is applied to the input circuits as usual. The conditioned input signal is not applied directly to the main gate in this mode. Instead, the conditioned output of the input circuit is applied to the control circuits, where it is used to form the gate-enable signal to the main gate. The normal output of the time base is selected and fed to the other input of the main gate. Now the bcd counters will be used to count the number of timing pulses coming from the time base. The input signal will be used to control the gating. The counters will repeatedly count and measure the number of time-base pulses that occur during the one-period gating intervals.

To make a period measurement, the control circuits first generate a reset pulse to the bcd counters. Then, a one-period-long gating signal derived from the input is applied to the main gate. During this time, the main gate is enabled, and timing pulses from the time base are passed through and accumulated by the bcd counters. A front-panel rotary or push-button switch is used to select the desired time-base frequency. This allows the user to control the measurement resolution. Assume that an input frequency of 2500 Hz is applied to the counter. This represents

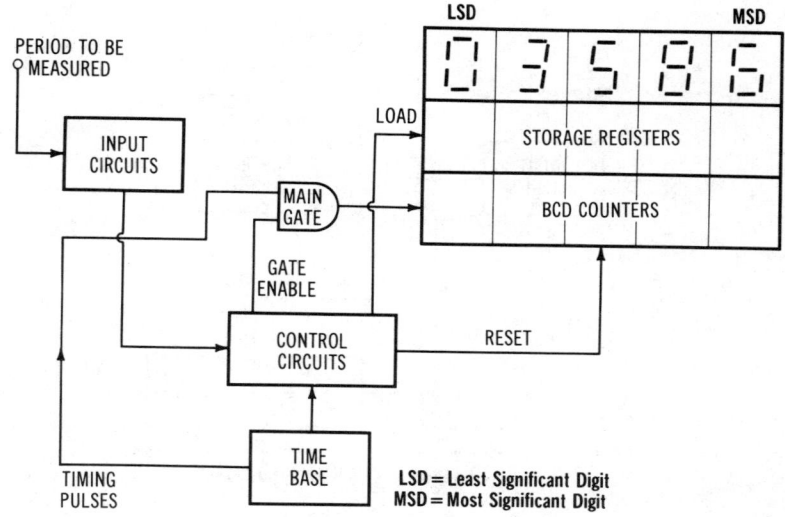

Fig. 2-18. Counter circuits connected for period measurement.

a period of 400 microseconds. If the 1-MHz time base output is selected, 1-microsecond timing pulses will be applied to the counter through the main gate. In one period of the input signal, 400 pulses will be counted; thus the display will indicate 400 microseconds.

A further improvement in the resolution of low-frequency measurements can be made by using a modified version of the period mode. Generally known as *period averaging,* this mode divides the input signal by a factor of 10, 100, or 1000, thereby extending the gating period by 10, 100, or 1000. The number of timing pulses counted during this extended interval is the sum of the pulses for the extra periods. Dividing this sum by the division factor gives the period average.

In some frequency counters, an extra set of decade frequency dividers is provided especially for this purpose. Fig. 2-19 shows how the extra decade counters are connected in the period-averaging mode. With SW1 in position A, the normal period mode is selected. Putting SW1 in position B selects the period-averaging mode. With this arrangement, the input signal can be divided by 10, 100, or 1000 by setting switch SW2 to the proper position. This allows the measurement to average 10, 100, or 1000 period intervals. The output of the input circuits is applied to the divider chain. The output of the dividers is then applied to the control circuit. The control circuit then generates the gate-enable signal to the main gate; this signal is 10, 100, or 1000 times the period of the input signal. As usual, the time-base frequency is selected to provide the desired measurement resolution.

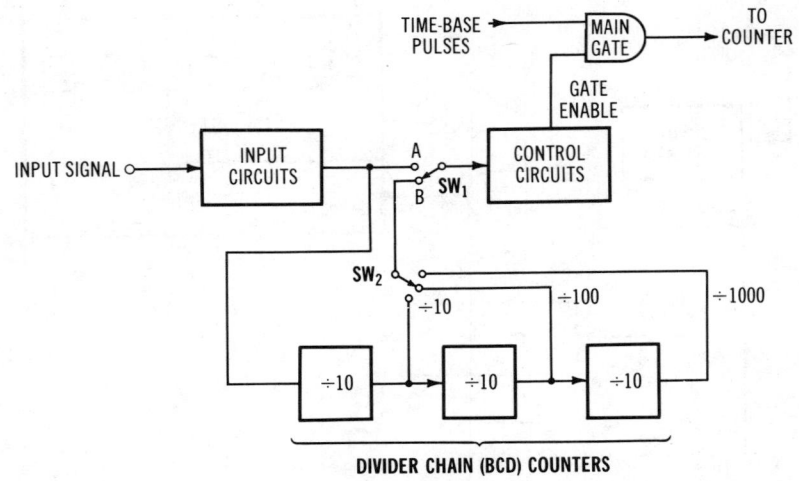

Fig. 2-19. Counter circuitry for period averaging

Depending on the number of periods being averaged, the display will read a value that is 10, 100, or 1000 times the true period. For that reason, to read the correct period, the display value must be divided by the appropriate factor. Since the factor is a power of 10, the correct period can be indicated by simply moving the decimal point one, two, or three places to the left to adjust for the division of the input signal.

The period-averaging mode is an excellent method of increasing the precision of measurement of low-frequency signals. Little or no additional circuitry is required to perform this measurement; the basic elements of the counter circuit are simply rearranged.

**The Totalizing Mode**

The totalizing mode is the simplest form of operation for a digital counter. In this application, the counter is simply used as a device for counting, totalizing, or accumulating the number of electrical events that occur at the input. In this mode, the internal time base is not used. Instead, the main gate is controlled by a front-panel switch or an external signal that will enable the main gate.

Fig. 2-20 shows a block diagram of a frequency counter used in the totalizing mode. The signals or events to be counted are applied to the regular counter input, where they are conditioned by the input circuits and applied to the main gate. The gate itself is enabled by the control circuits which, in turn, are operated by front-panel switches. A front-panel button is used to reset the decade counter chain manually. When this button is depressed, the decade counters are reset to zero and the display is cleared. Another front-panel button is used to enable

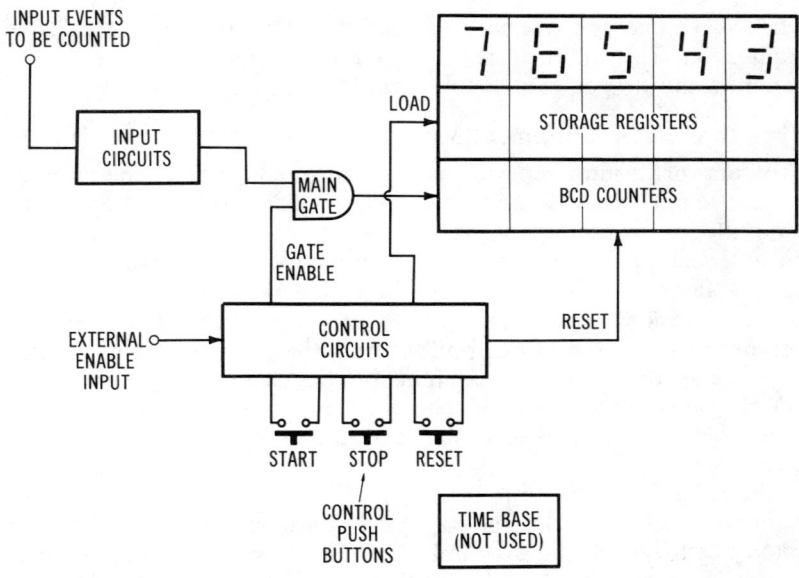

Fig. 2-20. Counter circuits connected for the totalizing mode.

the gate and start the count. Then as the electrical input pulses occur, they are counted and accumulated in the decade counter and displayed. This counting or accumulating of the input pulses continues indefinitely until the front-panel stop switch is pressed. When this switch is depressed, the control circuits inhibit the main gate. The counter stops totalizing and simply stores, or remembers, the count. The counter content is continuously displayed. If the gate is again enabled by the start button, the input pulses simply continue to add to, or accumulate to, the previous count.

An example of a totalizing-mode application is counting the number of devices moving along a conveyer belt. At one point along the belt, a light source and photocell are positioned so that the items to be counted (boxes, cans, widgets, etc.) pass between the light source and the photocell. When the light beam is interrupted by a passing item, the photocell generates an electrical pulse that is applied to the counter input. For each item that passes the photocell, an electrical pulse is generated and is counted by the counter. Many other applications require similar counting operations for which the totalizing mode is applicable.

In addition to the manual front-panel control switches used for controlling the gating function, an external signal may also be used. This pulse can be generated by some external circuit or device that knows when to enable and disable the main gate for the proper totalizing application. This input pulse is applied to the control circuits, which generate the proper gate-enable signal.

**Time-Interval Measurement Mode**

In some of the more sophisticated laboratory counters, a time-interval measurement mode is provided. This mode allows a variety of time-interval measurements to be made. For example, pulse width, pulse spacing, rise and fall times, and duty cycle can be measured. This mode is somewhat of a cross between the totalizing and period modes. It is like the period mode in that the control circuitry is arranged so that the decade counters count timing pulses from the time base. This is what provides the time-measurement function. For example, if the time-base 1-MHz signal were selected, then the 1-microsecond interval pulses would be counted, and the counter display would show the time in microseconds.

The main gate in the time-interval measurement is controlled by the external signal or signals to be measured. These external signals tell the gate when to be enabled and disabled. In this respect, the time-interval measurement mode is like the totalizing mode when an external input signal is used. The external signal or signals effectively tell the gate when to open and when to close.

Fig. 2-21 shows a block diagram of a digital counter with time-interval measurement capabilities. Note that this counter has two complete and separate input-signal conditioning circuits. Like a standard input circuit on a frequency counter, these circuits consist of attenuators, protection circuits, an amplifier, and a trigger circuit. Some means of adjusting the trigger level and slope (polarity) is also provided. This allows the user to select the point on the input waveform at which triggering takes place. The outputs of these input circuits are applied to the counter control circuits. The control circuits in turn generate the gate-enable pulse used to enable or disable the main gate.

To illustrate the time-interval measurement mode, let us consider a pulse-width measurement. For example, it may be desirable to determine the length (time duration) of a logic pulse. As shown in Fig. 2-22, the width, or time duration, of a logic pulse is measured at the 50% amplitude points on the leading and trailing edges of the pulse. This pulse is applied to one of the input-signal conditioning circuits. The other circuit is not used in this application. The trigger level is adjusted so that the main gate is switched at the 50% amplitude points on the input waveform. This means that when the input pulse rises to the 50% mark, the main gate is opened and time-base pulses are allowed to reach the counter. When the 50% level on the trailing edge of the pulse is reached, the gate is disabled, and the count is terminated. The number displayed on the counter is the duration of the pulse. If the counter reads

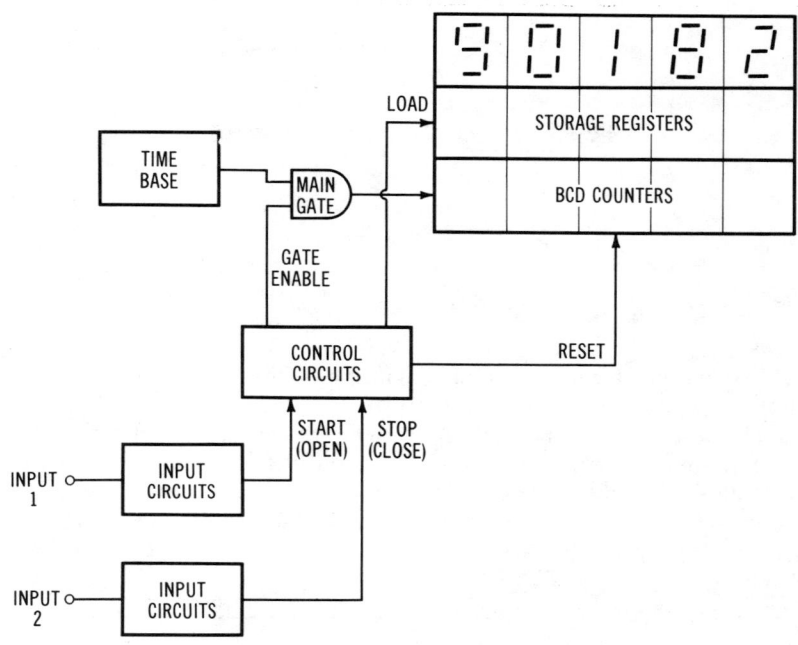

Fig. 2-21. Counter circuits connected for time-interval measurements.

734 and the 1-MHz time-base signal is selected, the pulse duration is 734 microseconds. Naturally, other time-base signals may be selected, depending on the actual pulse width and the desired resolution.

Another example of this mode shows how the input circuits are used to control the gate. Assume that it is desirable to measure the time lapse between the occurrence of two electrical signals, A and B (Fig. 2-23). The two separate signals are applied to the two independent inputs of the counter. The trigger levels on the input circuits are adjusted so that triggering occurs at the appropriate points on the input waveforms. The counter is initially reset. When the first input pulse (A) occurs, the main gate is opened, and time-base pulses are accumulated in the counter. When the other input signal (B) occurs, it disables the main gate, thus inhibiting the time-base pulses. The count displayed by the counter at this time represents the time interval between the pulses. The time-base frequency is selected to provide the best resolution.

A variation of the time-interval mode available on some laboratory counters is time-interval averaging. Like period averaging, this mode measures the time interval many times (10, 100, or 1000) and provides a true statistical average output. The results are a great reduction in the errors generally associated with time-interval measurement and an

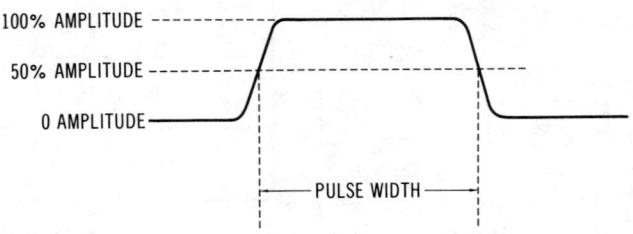

Fig. 2-22. Width of a pulse.

improvement in measurement resolution. This technique works only with repetitive or periodic signals.

The time-interval measurement mode is extremely versatile. With two input-signal conditioners whose amplitude and trigger polarity are selectable, a wide range of measurements can be made on pulse-type signals. Previously, such measurements had to be made with an oscilloscope. Even the highest-quality oscilloscopes provide at best only

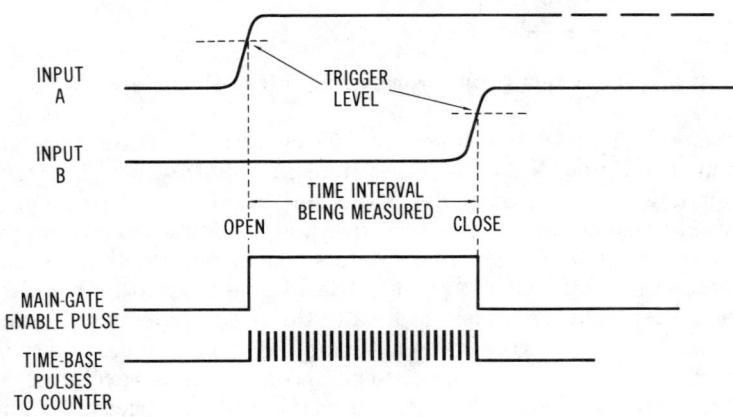

Fig. 2-23. Measuring the time interval between two input signals.

moderate measurement accuracy, since the measurement is dependent so much on the visual observations and manual adjustments made by the user. With an electronic counter, measurements such as pulse width, pulse spacing, and rise time can be made with an accuracy not possible before. Such measurements will be discussed in more detail later in this book.

CHAPTER **3**

# Counter Specifications and Error Sources

The specifications of a digital counter are those characteristics that define its limits of performance and its capabilities in measurement. Some typical counter characteristics are input-voltage sensitivity and frequency range. It is important to know the specifications of the counter you are using. This will allow you to apply the counter properly. It is important that the counter have necessary characteristics to provide the desired measurement result. By knowing the capabilities and limitations of the counter, you can determine whether the instrument is suitable for your application.

Knowing the specifications of the counter and what they mean provides further understanding of the operation of the counter. As indicated earlier in this book, a thorough knowledge of counter operation provides the best insurance against improper use of the instrument. A good knowledge of counter operation and specifications will allow you to get the most out of a given counter.

An understanding of counter specifications will also allow you to define and select a new counter. Given the task of selecting and purchasing a new counter, you would first determine the specifications of the counter required for your particular application. You would do this by thoroughly defining your application in terms of the types of signals to be measured and their characteristics. You must also define the amount of accuracy required in your measurement. Once your uses have been thoroughly outlined and the measurement requirements determined, you can prepare a set of minimum specifications. You can then select the lowest-cost counter for meeting these specifications. The result of such an approach provides the best performance for lowest cost.

Like counter specifications, error sources must also be thoroughly understood if the counter is to be properly applied. Since the counter is a measuring instrument, you must know the accuracy of the time and frequency measurements. This can only be determined if you possess a good understanding of the sources of error in the counter. Even the most sophisticated and expensive digital counters are not error-free. Yet errors have been reduced to a very low level, thus permitting highly accurate time and frequency measurements to be made. Again, a knowledge of the error sources will permit you to obtain the best measurement possible.

This chapter gives a complete discussion of the specifications and error sources in modern digital counters.

## PERFORMANCE SPECIFICATIONS

There are several important specifications that fully define the capabilities and limitations of a counter. This chapter provides a thorough discussion of the most important of these specifications.

### Sensitivity

The sensitivity of a counter is the minimum input signal level with which the counter can be used. This is usually the smallest signal amplitude capable of reliably triggering the internal circuits. Typically, the sensitivity is expressed as a voltage value. It is most often expressed as the minimum rms value of a sinusoidal input signal.

Most modern digital counters have a sensitivity in the 10- to 100-millivolt range. A lower voltage value indicates better sensitivity. A counter whose circuits will trigger reliably at 20 millivolts is more sensitive than one that triggers at 90 millivolts.

The sensitivity usually varies over the frequency range of the counter. The best sensitivity is obtained over the midfrequency range. At the lower and upper frequency limits, the sensitivity decreases, meaning that it takes more input voltage to trigger the counter reliably. Fig. 3-1 shows a sensitivity versus frequency curve for a typical counter.

As a general rule, the more sensitive a counter is, the better a measuring instrument it is. However, it is important to note that high sensitivity is undesirable in some applications. This is particularly true if the signal being measured is noisy. Low-level, higher-frequency noise pulses superimposed on a lower-frequency signal to be measured can cause false triggering of the counter and thus produce inaccurate readings. In such applications, it is generally desirable to use a counter with less sensitivity. In some counters, sensitivity is adjustable with a front-panel control that allows you to select the sensitivity that is best for your application.

The sensitivity of a counter is a direct function of the characteristics of

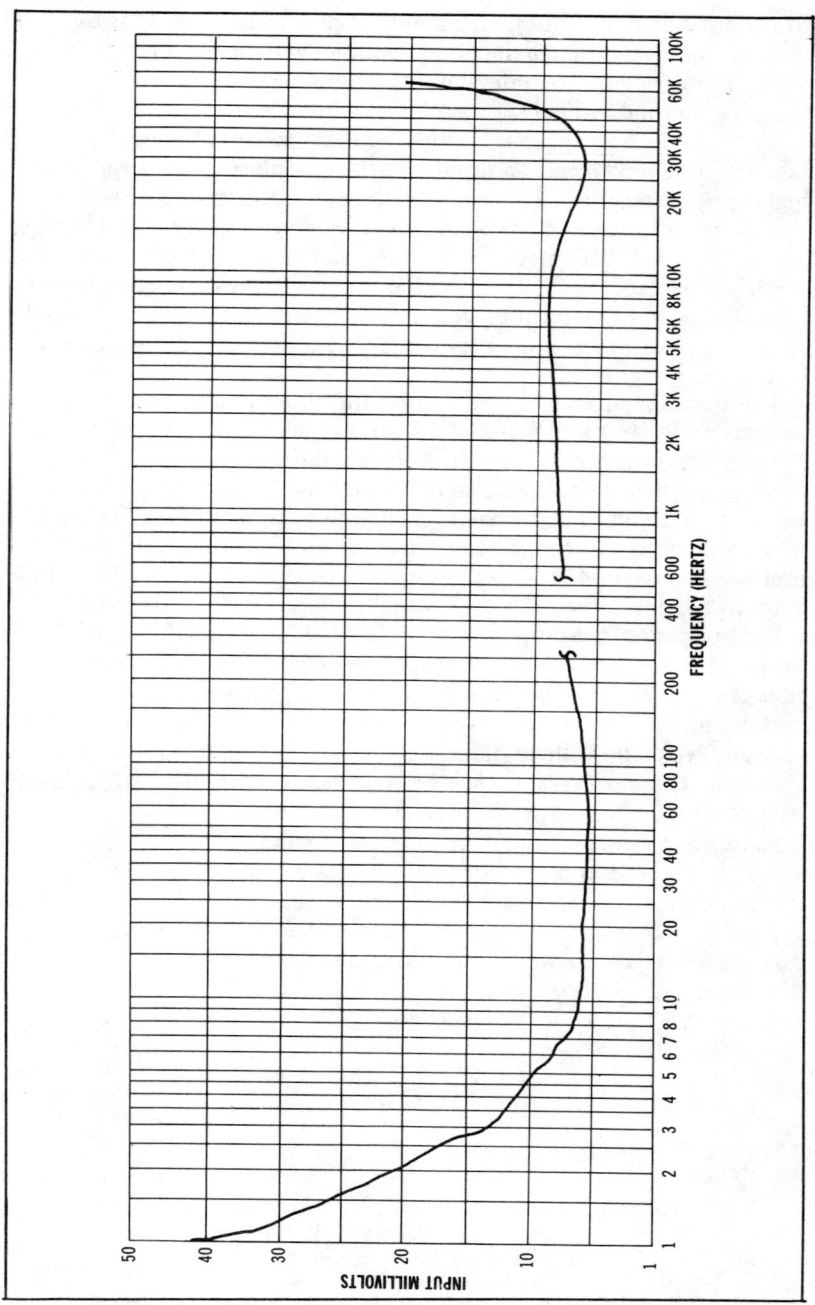

Fig. 3-1. Typical input sensitivity curve of a frequency counter.

the input-signal conditioning circuits. Specifically, the sensitivity is determined by the gain of the input amplifier and the hysteresis levels of the Schmitt trigger. To understand this better, we will consider each of these factors individually.

The Schmitt trigger circuit in the input-signal conditioner is a circuit that effectively converts an input signal, regardless of its shape, into a high-speed logic pulse. The output of the Schmitt trigger is a logic signal that switches between two voltage levels, defined as logic 0 and logic 1, which are compatible with the main gate. The switching between the two logic levels occurs at a very high rate of speed, meaning that the rise and fall times of the output signal of the Schmitt trigger are extremely short. In most modern counters, these rise and fall times are in the nanosecond region.

The switching from one logic level to the other occurs at specific input voltage levels. In some Schmitt trigger circuits these voltage levels are fixed by the design of the circuit. For example, a typical Schmitt trigger circuit may switch from zero to one when the input voltage exceeds 1 volt. The Schmitt trigger may then switch from one to zero when the input voltage drops to 0.5 volt. Fig. 3-2 shows an example. Here an input sine wave of 4 volts peak-to-peak is applied to a Type 7413 TTL Schmitt trigger. When the positive-going input threshold $(+V_T)$ of 1.7 volts is exceeded, the output switches from 1 to 0. When the signal drops below the negative-going threshold $(-V_T)$ of 0.8 volt, the output changes from 0 to 1. The most important point to note here is that the switching between the two output logic levels does not occur at the same voltage level. The voltage difference between these two input levels is known as the *hysteresis* or *hysteresis window*. For the 7413 Schmitt trigger, the hysteresis is $1.7 - 0.8 = 0.9$ volt.

In some types of trigger circuits, the voltage triggering level is adjustable. For example, when an IC comparator circuit is used to shape the input signal, the trigger level can be set by varying the voltage applied to the comparator reference input. Refer to Fig. 3-3. A voltage divider made up of $R_1$, $R_2$, and $R_3$ connected between positive and

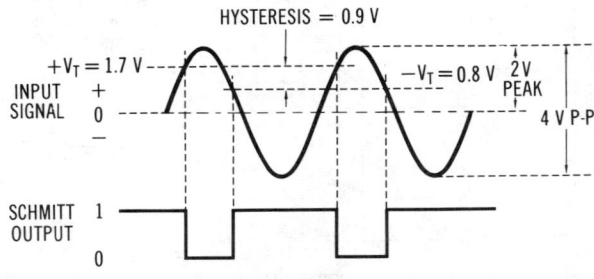

Fig. 3-2. Input and output waveforms of a typical Schmitt trigger.

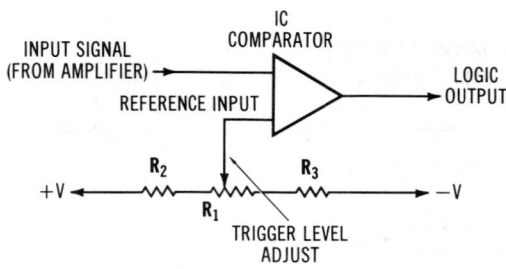

**Fig. 3-3. An IC comparator used as an input trigger circuit.**

negative voltage sources $+V$ and $-V$ provides a variable voltage to the reference input. By means of $R_1$, the user selects the voltage trigger level at which the output switches from one state to another. This voltage may be either positive or negative depending on the setting of $R_1$. With $R_1$ at the center of its rotation, the reference input is zero volts. Regardless of whether the input signal is positive, negative, or zero, whenever it goes above or drops below the reference level, the output switches. Keep in mind that even IC comparators exhibit the characteristic of hysteresis. The hysteresis can be set to almost any desired level depending on comparator circuit configuration. However, it is usually set to a low level. Values from 10 millivolts to 1 volt are typical, depending on the input levels expected and the amount of amplification ahead of the comparator.

An input amplifier is often used to raise low-level signals to a high enough value to operate the trigger circuit reliably. Considering the amplifier and trigger circuits as one, the true hysteresis is the hysteresis of the trigger circuit divided by the gain of the amplifier. For example, if the hysteresis is 0.8 volt and the amplifier gain is 100, the overall hysteresis is $0.8/100 = 8$ millivolts. This overall hysteresis is the sensitivity of the counter. You can think of the separation between the hysteresis levels as the peak-to-peak sensitivity of the counter. Usually the sensitivity is expressed in rms values. Therefore, the sensitivity is the rms value of the peak-to-peak difference between amplified hysteresis levels.

### Input Impedance

The input impedance of the counter is the impedance looking into the input circuit of the counter. Input impedance is the load that the circuit or device whose output is being measured sees. The input impedance of a counter is not infinite, and thus it does cause some loading of the circuit or device whose output is being measured. The input impedance of the counter appears in shunt with the measured signal. However, the input impedance of most counters is high enough to minimize these loading effects.

Most commercial counters have an input impedance that is a 1 megohm resistance in parallel with a small value of capacitance in the range 5 to 100 picofarads. The resistance value is typically fixed by the input attenuator resistors. The capacitance is the result of any frequency-compensating capacitors used in the attenuator plus any stray or shunt capacitance in the instrument. The capacitance of the coaxial cable used to connect the counter to the signal source must also be considered. See Fig. 3-4.

At low frequencies (below 1 MHz), the input capacitance has negligible effect. But at higher frequencies (1 MHz and above), the input capacitance becomes an important part of the input impedance. The reactance of the input capacitance begins to provide some shunting effect at these frequencies. Nevertheless, the 1-megohm input impedance is usable at frequencies up to approximately 100 MHz. At this point, the effect of the shunt capacitance becomes too great for reliable measurements.

At frequencies above 100 MHz, most counters are set up for a 50-ohm input impedance. Signals in this frequency range are radio signals which are generated at lower inpedance levels. Therefore, this low-impedance input does not cause excessive loading. At lower impedance levels, the effect of any shunt capacitance is extremely small. In fact, the 50-ohm input circuits are usable well into the microwave frequency range. In most low-frequency counters, the 1-megohm input impedance is standard. In general-purpose high-frequency counters, both 1-megohm and 50-ohm inputs are provided. Some microwave counters have only a 50-ohm input.

**Frequency Range**

The frequency range is the bandwidth of the counter. This is best defined by the upper and lower frequency limits between which the counter operates reliably. Circuitry limitations define the upper and lower frequency limits of the counter. The input circuit, the main gate, and the input decimal counter all have upper frequency limits.

Most counters have a lower frequency limit of approximately 5 Hz. This low-frequency limitation is due primarily to the ac coupling capacitor used at the input ($C_1$ in Fig. 3-4). The value of this capacitor combined with the 1-megohm input impedance defines the low-frequency cut-off of the counter. See Reference VI.

The upper frequency limit of the counter varies widely depending on the type and quality of the counter. Most modern counters have an upper frequency limit of at least 30 MHz. This upper frequency limit is determined primarily by the frequency response of the input amplifier circuit, the propagation delay of the trigger circuit and the main gate, and the upper frequency-counting limit of the *input* decade counter. Today it is not unusual for even basic low-cost counters to be able to

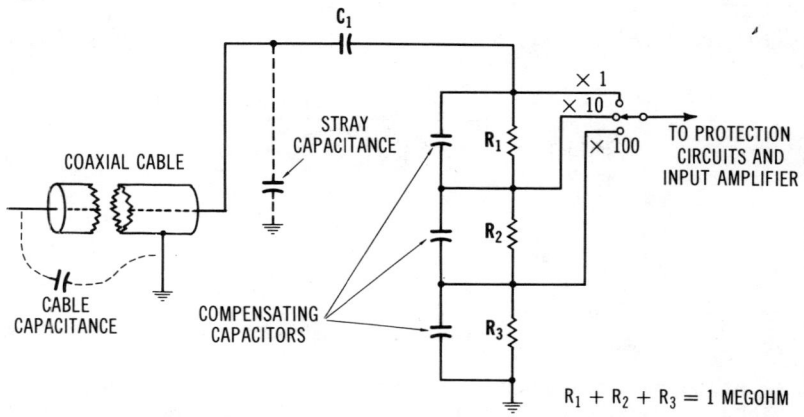

Fig. 3-4. Factors determining counter input impedance.

## REFERENCE VI

## Cutoff Frequency and Bandwidth

The bandwidth of an electronic circuit is the difference between the upper and lower cutoff frequencies, $f_2$ and $f_1$, as shown in Fig. VI-1. These cutoff frequencies occur as the result of equivalent low-pass and high-pass RC networks in the circuit. For example, the low-frequency cutoff is determined by the high-pass RC filter circuit shown in Fig. VI-2. The values of R and C determine the cutoff frequency. Below this frequency, attenuation of the input signal occurs rapidly. Above this frequency, the signal passes without attenuation.

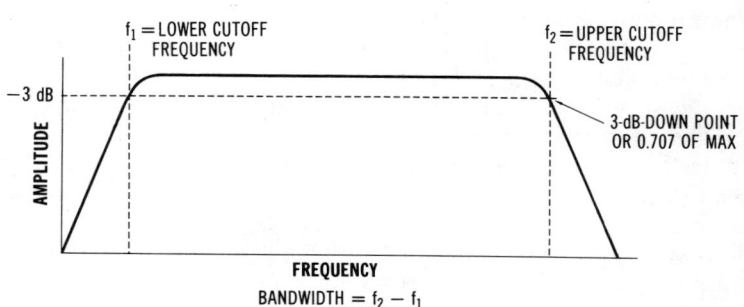

Fig. VI-1. Bandwidth and frequency response of an electronic circuit.

*Continued on next page.*

REFERENCE VI *(cont.)*

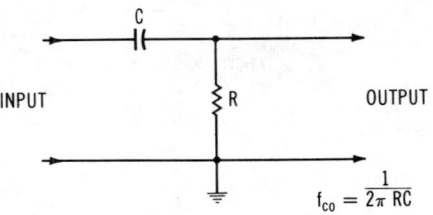

Fig. VI-2. High-pass RC network.

At the cutoff frequency, the attenuation is 3 dB. At this point, the voltage amplitude of the output is 0.707 of the input voltage. The cutoff frequency ($f_{co}$) can be computed with the formula

$$f_{co} = \frac{1}{2\pi RC}$$

The value of R is in ohms, and C is in farads. Pi ($\pi$) is 3.1416. Assume an input impedance (R) of 1 megohm and a capacitance of 0.02 µF. The low-frequency cutoff is

$$f_{co} = \frac{1}{2(3.1416)\,(1{,}000{,}000)\,(0.02 \times 10^{-6})}$$
$$f_{co} = 7.96 \text{ Hz}$$

measure frequencies as high as 100 MHz. Special high-frequency counters are also available to measure frequencies as high as 40 GHz.

**Resolution**

The number of digits in the counter output display determines the resolution of the time or frequency measurement. Keep in mind that resolution is also a function of the gate time or time base selected for the given measurement. Remember, too, that the greater the resolution, the higher the precision of the measurement will be. Most modern counters have at least five digits in the readout. Counters used for measuring high frequencies up to the gigahertz range can have as many as twelve display digits.

Resolution is the degree to which one can distinguish the fineness of detail in a measurement. It is also a measure of the smallest possible increment of change. In most modern counters where the time base can be selected for a given time or frequency measurement, the resolution is variable. Resolution defines the smallest increment displayed. In other words, the least significant digit of the display represents one unit of resolution as determined by the mode and the time-base setting.

## Dynamic Range

Dynamic range refers to the range of input signals that may be applied to the amplifier and input-conditioning circuitry. More specifically, the dynamic range refers to the limits of the input over which the input amplifier remains linear. The linear operating range of an amplifier refers to the range of input signals over which little or no distortion of the input signal occurs. The dynamic range, therefore, is the difference between the upper and lower voltage levels of input signals between which the amplifier remains linear and reliably triggers the Schmitt trigger circuit. The lower limit of the dynamic range is typically the sensitivity level of the amplifier. This is usually in the millivolt region. The upper voltage level of the dynamic range is the maximum input voltage that when attenuated to the maximum degree by the input attenuator is below the clamping level of the input protection circuits.

Most input-signal conditioning amplifiers remain linear and do not produce distortion of the input signal if operated well below the maximum input voltage level. It is only when the input voltage level becomes excessive that distortion begins to occur. If the input signal is too large and the attenuator is unable to reduce that signal level to a point below which the voltage-protection clamping circuits work, non-linear operation will occur, and distortion of the input signal will result.

In the frequency-measurement mode, nonlinear operation is not particularly detrimental to performance. In fact, the input signal is ultimately shaped by the Schmitt trigger circuit into a nonlinear pulse signal. However, in the time-interval measurement mode, it is desirable to preserve the rise and fall times of pulses so that accurate measurements of rise time, pulse width, and pulse spacing can be made. If distortion occurs or the bandwidth of the amplifier limits the frequency range, this timing will be destroyed. The result, of course, is inaccuracy in measurement.

## Maximum Input Voltage

The maximum input voltage is what the term implies. It is the maximum amount of input voltage that may be applied to the input without damaging the input circuits or degrading performance. Fig. 3-5 shows a plot of the maximum input voltage of a typical modern counter. Note that the maximum input voltage varies considerably with the input frequency. The higher the frequency, the lower the maximum input voltage becomes for best precision of measurement.

## Time-Base Characteristics

Perhaps the most important specifications of a digital counter are those of the time base. The accuracy of time and frequency measurements in a counter is directly related to the quality of the time base. In frequency measurements, for example, the time-base oscillator

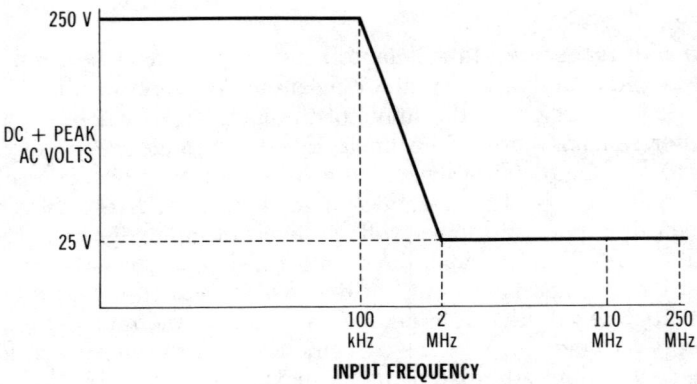

Fig. 3-5. Input derating curve for a counter with an upper frequency limit of 250 MHz.

must generate timing signals whose frequency is extremely accurate in order that the gating pulse applied to the main gate is precisely one second or some multiple or submultiple of one second. Any deviation from the exact value will introduce measurement errors. In time-interval measurements, the time base must accurately generate pulses of the correct frequency in order to produce an accurate measurement of time interval. While digital measuring techniques provide high resolution and the convenience of direct decimal display, it is time-base accuracy that ensures the validity of the measurement. Reference VII defines accuracy.

Most time-base circuits use quartz-crystal oscillators. The important characteristics of the time base are the actual frequency of the crystal and the stability of its frequency. The ability of the oscillator to maintain a constant frequency in spite of changes in ambient temperature and supply voltages gives an indication as to the quality of the time base.

Most time-base oscillators use a crystal with a frequency of 10 MHz. Other crystal frequencies occasionally used are 4, 2, and 1 MHz. The exact frequency of the time-base oscillator is not critical as long as frequency dividers are provided to generate the necessary gating pulses for frequency measurement. For time-interval measurements, the oscillator frequency is important. The higher the time-base frequency, the greater is the resolution obtained in period and time-interval measurements. For example, with a 10-MHz crystal oscillator, pulse intervals of 100 nanoseconds can be generated. With a time base of 1 MHz, one-microsecond pulses will be generated. Naturally, a time measurement made with 100-nanosecond pulses gives better resolution than one made with 1-microsecond pulses.

The number of time-base signals available to the user is also an important consideration. Most time bases incorporate a chain of decade

> ## REFERENCE VII
>
> ### Accuracy
>
> Accuracy is a number that indicates how close a calculation or measurement is to its true value. Accuracy is the ratio of the actual value to the true value, expressed as a percentage. The following formulas are used to calculate accuracy.
>
> $$\text{Accuracy} = \frac{\text{Calculated value}}{\text{True value}} \times 100\%$$
>
> Use this formula if the calculated value is less than the true value. If the calculated value is greater than the true value, use the formula
>
> $$\text{Accuracy} = \frac{\text{True Value}}{\text{Calculated value}} \times 100\%$$
>
> For example, suppose the time base in a counter is not perfectly accurate. It measures a true frequency of 1.5 MHz as 1.489 MHz. The accuracy is
>
> $$\text{Accuracy} = \frac{1.489}{1.5} \times 100 = 99.267\%$$

dividers after the crystal oscillator to generate decade steps of timing pulses. In many general-purpose counters, all of the various time-base signals are made available so that the user may select the one best for his measurement. In other counters, only one or two time-base signals may be made available.

Most time-base circuits are adjustable. This allows the user to adjust the time base to an exact frequency by comparing the time-base signal with an accurate frequency standard. Once the time base has been adjusted or calibrated, the quality of the measurements from this point on depends upon how stable the oscillator is. Important factors affecting the oscillator frequency are temperature, short-term stability, long-term drift, aging, and power-supply voltage changes.

The factor that most influences the frequency of the crystal oscillator is the ambient temperature. The crystal frequency varies considerably with changes in temperature. When a counter is first turned on, the crystal is cold. However, due to the heat of the surrounding electronic circuits, the crystal temperature will rise. Even a small change in temperature will cause the crystal frequency to vary. Once the unit has been on for 30 minutes or more, the temperature of the crystal will essentially stabilize, as will the time-base frequency. However, even at this point further changes in frequency can be encountered if the temperature changes. The frequency can either increase or decrease with temperature, depending on the angle of cut of the crystal. By choosing the proper angle of cut, frequency variation with temperature can be minimized.

Frequency variation due to temperature can be controlled and reduced. The most common way of doing this is to design the oscillator circuit so that it is temperature compensated. Most quality counters use temperature-compensated crystal oscillators (tcxo) in the time base. A temperature-compensated crystal oscillator incorporates a temperature sensor that causes a variation in voltage or current to control another component, such as a voltage-variable capacitor, to adjust the frequency of the oscillator to correct for changes caused by temperature variations. With such an arrangement, the frequency stability and hence measurement accuracy become much better than with a conventional uncompensated crystal oscillator.

A further improvement in temperature stability is achieved by using a crystal oven. With this arrangement, the crystal oscillator is housed in a sealed oven whose temperature is controlled with a built-in heating element. In the simplest form of oven, the heating element is turned off or on depending on the temperature. A temperature sensor determines the temperature of the housing and turns the heating element on when the temperature drops below a certain level. When the temperature reaches a specified upper level, the sensor turns off the heating element, thus allowing the crystal to cool. This off-on control allows the crystal temperature to stabilize within a narrow range.

More sophisticated crystal ovens use a proportional control arrangement. Here, the heating-element supply voltage is made proportional to the difference between the inside and outside temperatures of the oven. Therefore, the heat is varied proportionally or continuously rather than having the heating element turned off and on. This reduces the cycling action of the oven as the heater is turned off and on and narrows the temperature variation of the crystal.

Very high-quality time-base oscillators often employ two ovens, one inside the other. By using proportional control on such ovens, the temperature of the crystal can be maintained within 0.01%. An extremely stable frequency is produced.

The frequency variation of a time base is typically expressed in terms of parts per million (ppm). See Reference VIII. In uncompensated crystal oscillator circuits, the frequency stability will usually be in the range of 5 to 10 ppm as the temperature varies from approximately 10°C to 40°C. The frequency variation in tcxo's is considerably less and can be in the 1-ppm range over a temperature variation of 0°C to 50°C. By using an oven, frequency variation can be reduced to well below the 1-ppm level over a wide temperature range.

Long-term drift of the time-base frequency is due primarily to crystal aging. As a quartz crystal is used, various internal changes take place to

> ## REFERENCE VIII
>
> ### Parts per Million
>
> Parts per million (ppm) is a term used to express the accuracy and stability of a measurement or circuit. This term designates the amount of variation between the true and measured values or the difference between the desired and actual values. One part per million simply means $1/1,000,000$, or $1 \times 10^{-6}$. In terms of a percentage, 1 ppm is 0.0001%. Five parts per million is $5 \times 10^{-6}$, or 0.0005%. If a 10-MHz time-base oscillator has a stability of 5 ppm, its frequency can vary as much as 0.0005%, or $0.000005 (10,000,000) = 50$ Hz, above or below 10 MHz.

cause its frequency of oscillation to vary slightly. For high-quality counters, an aging rate of less than 1 ppm per month or longer is typical.

The long-term stability due to aging is not a serious problem. With most counters, it is negligible as far as measurement accuracy is concerned. However, it should be borne in mind that this aging effect does occur. It can be easily corrected for by recalibrating the time base at frequent intervals. If the counter is used considerably, recalibration should be more often than with counters used less frequently. The reason for this is that aging occurs only while the crystal is being used.

Short-term stability refers to the ability of the crystal oscillator to maintain a constant frequency over a short period of time. Because of various crystal imperfections and slight instabilities in oscillator circuits, there are short-term variations in the oscillator frequency. Fig. 3-6 illustrates these minor fluctuations in frequency. Note how these short-term changes are superimposed on a slower-changing age-rate curve.

The remaining factor that influences the stability of the time-base oscillator is power-supply voltage. Changes in power-supply voltage applied to the oscillator circuit will cause frequency variations. These variations in supply voltage are typically caused by variation in the ac line-voltage input to the power supply. They can also be caused by variations in the load on the power supply in the instrument itself. Typically, however, this load is constant or nearly constant, and most voltage variations are a function of ac line variations. The effects of these ac line variations may be effectively eliminated with good power-supply regulation. Most high-quality counters use quality voltage regulators that effectively reduce frequency changes due to voltage variations to a very low level. In a counter with good regulation, the frequency deviation is less and $1 \times 10^{-10}$.

Temperature changes, aging, and power-supply variation in the time base all introduce measurement error. Other sources of error also result

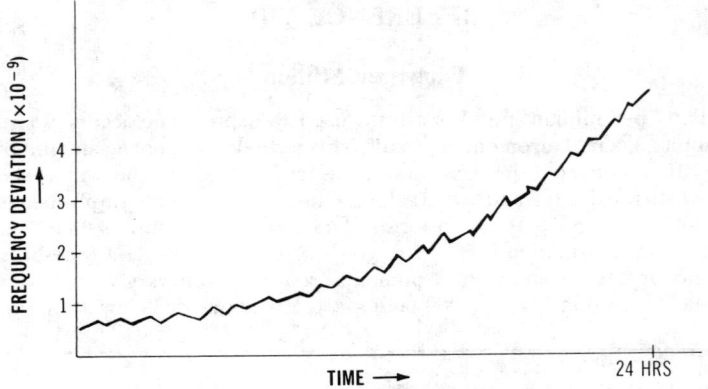

Fig. 3-6. Long- and short-term stability of time-base crystal frequency.

in inaccurate measurements. But time-base error is by far the most serious. A good time base is essential if the counter is to be relied upon to make accurate measurements. See Reference IX for a discussion of error. Error sources other than time-base error are covered in the next section.

## ERROR SOURCES

It was indicated earlier that the quality of measurement in a digital counter is determined by the resolution and accuracy of the instrument. The greater the number of digits and decade counters, the higher the measurement resolution will be. High resolution is desirable, but it is no guarantee that the measurement will be accurate. Accuracy, on the other hand, is primarily a function of the quality of the time base. An accurate and stable time base will produce highly accurate measurements of both time and frequency. Time-base error has already been discussed. However, there are other factors that enter into the quality of the measurement. Specifically, a variety of error sources can greatly affect the measurement accuracy. These include trigger errors and an inherent ±1 count error that is characteristic of all digital measuring instruments. It is important for you to understand these error sources so that you can compensate for them or otherwise minimize them to ensure high measurement accuracy.

### The ±1-Count Error

The ±1-count error arises because the signal being measured occurs at random with respect to the gating signal created by the time base. The

## REFERENCE IX

### Error

Error is a number that indicates the amount by which the result of a calculation or measurement differs from the true or correct value. Error is the amount of inaccuracy in the result. It is usually expressed as percentage, calculated by the formula

$$\text{Error} = \frac{\text{True value} - \text{Calculated value}}{\text{True value}} \times 100\%$$

If the calculated value is greater than the true value, simply reverse the positions of the true and calculated values in the numerator of the formula; this will eliminate a negative result. Naturally, the smaller the error, the higher the accuracy and the closer the calculated value is to the true value.

For example, suppose that a counter measures a frequency of 6.98 MHz. You know that the true value is 7 MHz. The error is

$$\text{Error} = \frac{7 - 6.98}{7} \times 100 = \frac{0.02}{7} \times 100 = \frac{2}{7} = 0.2857\%$$

Another method of calculating error is to subtract the percent accuracy from 100%.

$$\% \text{ Error} = 100 - \% \text{ Accuracy}$$

For example, the error in the example in Reference VII is

$$\text{Error} = 100 - 99.267 = 0.733\%$$

---

signal to be measured and the gating pulse are not synchronized. The decade counters may actually register one more or one less count than the actual number of counts because of this lack of synchronization. The unsynchronized nature of the two signals is often referred to as *noncoherence*. The resulting ±1-count difference is also referred to as the *quantization error*.

Fig. 3-7 illustrates the count ambiguity that can be caused by the nonsynchronous nature of the pulses to be counted and the gating signal. The input signal is applied to a frequency counter, conditioned by the input circuits, then applied to the main gate. The input bcd counter is triggered by the trailing (negative-going) edge of the input signal. In this example, Gate 1 is a one-second gating signal. Determine the number of pulses (trailing edges) that occur during this one-second interval. Note that 7 pulses are counted. Gate 2 is another one-second pulse, but it has a different time position with respect to the input. Again determine the number of input pulses that occur. The counter will count 8 pulses with Gate 2. The frequency of the input signal is constant, as is

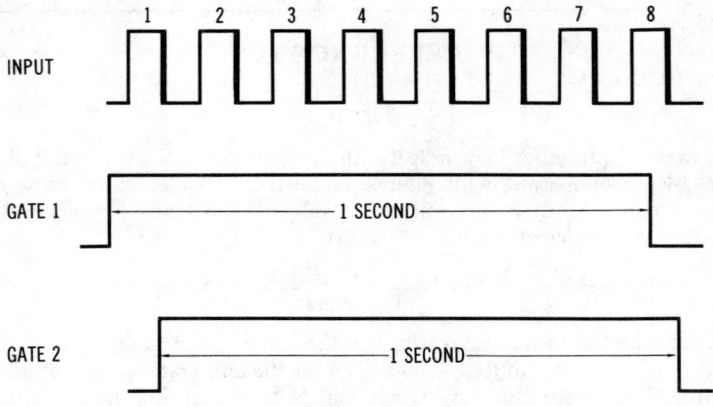

Fig. 3-7. Illustration of ±1-count error.

the gate-pulse duration. It is simply the timing relationship between the two that causes the false count.

The percentage of error based on the number of pulses counted during the gating interval can be determined by the expression

$$\% \pm 1\text{-count error} = \frac{1}{\text{number of counts}} \times 100$$

In the example of Fig. 3-7, the error is $\frac{1}{8} \times 100 = 12.5\%$.

This error can also be illustrated graphically as shown in Fig. 3-8. For frequency measurements, the error becomes extremely small at the very high frequencies, but at low frequencies the error is significant.

For time-interval measurements, the error is small when the time-base frequency to be counted is high with respect to the

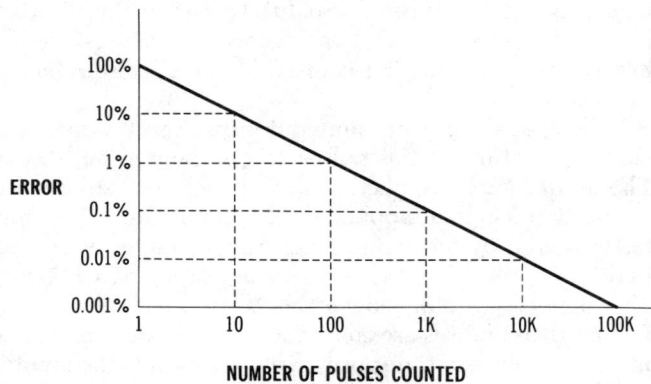

Fig. 3-8. Percent error due to ±1 count as a function of the number of pulses counted.

gating-pulse width. However, when the period of the time base is approximately the same order of magnitude as the gating-pulse width, a major error occurs.

There is little or nothing that can be done about the ±1-count error. As long as you are aware of it, you can make adjustments in your measurement technique to compensate for it. For example, do not use direct counting methods for measuring low-frequency signals. Instead, use the period-measurement mode to improve the accuracy. When the period-measurement mode is used for measuring low frequencies, the number of counts per period can be extremely large, particularly if the period-averaging mode is used. Because of the very high number of counts, the ±1-count error becomes less significant.

**Trigger Error**

The trigger error is a counting error that results from noise on the input signal. Many signals whose frequency or period is to be measured will be accompanied by noise or transients which can affect the measurement. Low-amplitude noise typically will not cause measurement error. But if the noise tends to become large in amplitude with respect to the measured signal, serious errors can result.

Trigger error due to noise in the frequency-measuring mode is not a significant problem. If the noise is not large enough to cause false triggering, no error will be introduced into the measurement. This error is effectively absorbed by the ±1-count error. This is illustrated in Fig. 3-9A. Here the amplitude of the noise is less than the counter sensitivity/hysteresis level, so no gross errors occur. Of course, there is a variation in where the actual trigger occurs because of the noise, but this irregularity is averaged out over many cycles.

However, should the noise be high in amplitude compared to the input signal, the error could be significant. If the trigger level is set so that the noise pulses are recognized, the Schmitt trigger may generate more output pulses than the actual number of input pulses to be measured. This is illustrated in Fig. 3-9B. The result, of course, is a frequency different from the actual frequency. Normally, this would be noticeable because of the erratic frequency reading that will be obtained under such conditions. Because the noise is random, the reading will not be steady or consistent. If this is the case, steps must be taken to reduce the level of the noise or readjust the triggering point to eliminate the false counts. Ordinarily, by adjusting the signal level or the triggering level, the noise amplitude can be made less than the trigger-voltage window on the Schmitt trigger, thus eliminating or at least greatly minimizing the error.

In time-interval measurements, the noise problem is altogether different. Keep in mind that in time-interval or period measurements, it is the input signal rather than the time base that generates the main gate

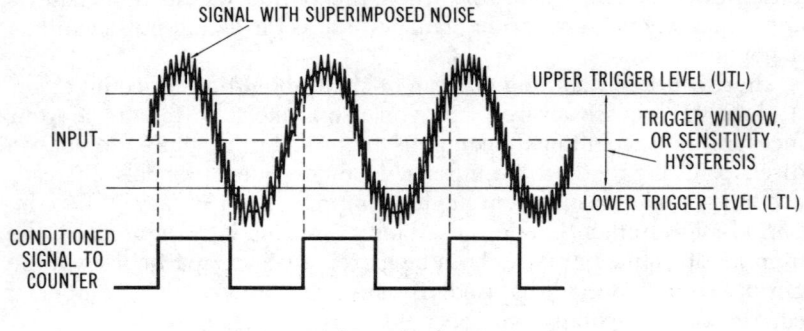

(A) *Low-level noise.*

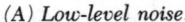

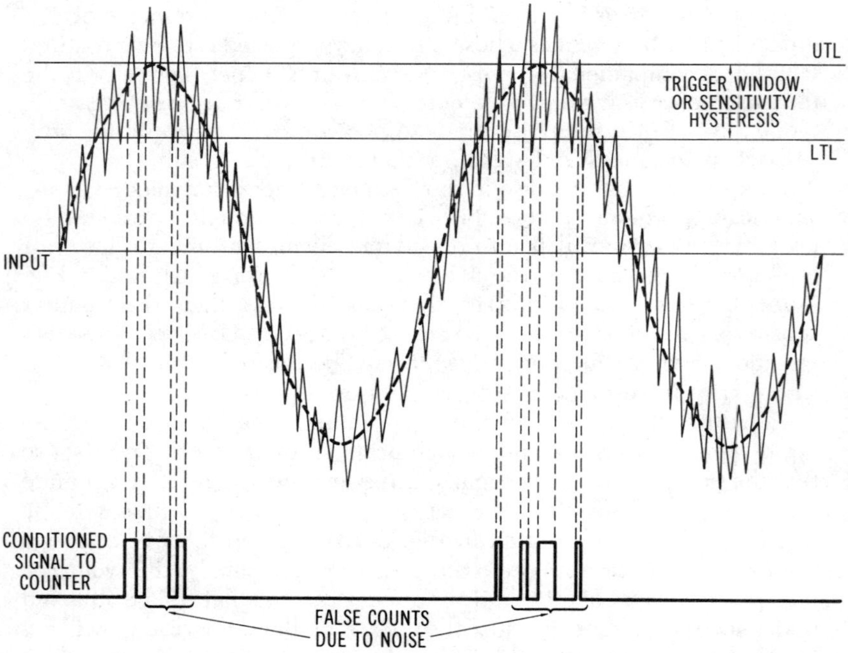

Fig. 3-9. Effect of noise on counting.

signal. In other words, the input signal determines the gate time during which the time-base pulses are allowed to reach the counter. Noise on the input signal will definitely determine when the gate is opened and closed. The trigger error is randomly caused by the noise, and the effect is to cause the main gate to open too soon or too late and thus create an incorrect period of gating time.

This trigger error can often be reduced by using period averaging and multiple time-interval measurement modes. Instead of measuring for just one period or time interval, some multiple decade number of periods or time intervals are measured. This averages out the error and produces a more accurate measurement.

Triggering errors can cause all kinds of unusual and unstable readings. Depending on the amplitude of the input signal, the amplitude of the noise, the characteristics of the transients superimposed on the signal, and the trigger-level setting, a variety of conditions can exist. Erratic display readings are the usual indication of noise. This can be verified quickly by connecting an oscilloscope to the input. Typically, by adjusting either the input-signal amplitude or the trigger level, improved triggering conditions can be selected for maximum measurement accuracy.

A feature sometimes encountered in higher-quality counters is input filters. These filters are used to remove the noise from the input signal. They are a special part of the input-signal conditioning circuitry and can be switched into or out of the circuit, depending on the measurement conditions. These filters are often frequency selectable to eliminate noise at different frequencies.

A few sophisticated counters also provide special circuits for improving the measurement accuracy in some noisy applications. For example, one feature often incorporated is known as *time-interval hold-off* (TI hold-off) or trigger masking. In counters where the time interval of some external signal is being measured, the TI hold-off feature can be used to mask out noisy conditions temporarily.

Consider the noisy input signal shown in Fig. 3-10. The objective is to get an accurate measurement of the pulse width or time interval. The noise pulses at the beginning of the pulse are typical of those caused by contact bounce inherent in most mechanical switches or relays when they are closed. It is difficult to get an accurate measurement of pulse width under such conditions. The noise pulses will effectively generate trigger signals which thus result in wild variations in the display reading. However, by using the TI hold-off feature, an accurate measurement can be obtained. The TI hold-off is a special pulse that keeps the main gate enabled for a short period of time, thereby effectively masking the noise. Usually this period of time is adjustable with a front-panel control. The timing measurement will begin as the very first input pulse is generated. This also turns on the TI hold-off pulse, which effectively blocks out the noise pulses. The result is an accurate time-interval measurement.

## Miscellaneous Error Sources

There are other sources of error that can affect the accuracy of measurement in a digital counter. Most of these additional sources are

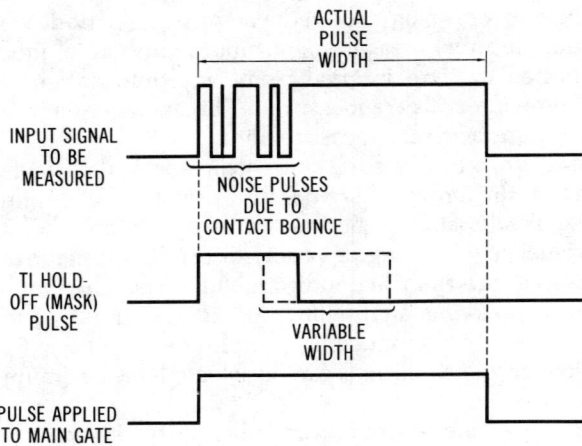

Fig. 3-10. Use of the TI hold-off feature to mask noise.

meaningful only in time-interval measurements. For example, when time-interval measurements are being made with two input channels used for generating the start and stop signals, any mismatch between the two channels will produce an error. Differences in frequency response, rise times, and propagation delays and other factors such as unequal probes, characteristics, or cable lengths can introduce significant timing errors. These are sometimes referred to as *systematic errors*.

CHAPTER **4**

# High- and Low-Frequency Counters

Virtually all digital counters, even the low-cost units, are capable of measuring frequencies well into the high radio-frequency range. Most common counters can measure frequencies up to 30 to 80 MHz. And, counters to measure frequencies significantly higher are readily available. In fact, with the counting techniques described previously, counters can be used to measure frequencies up to about 500 MHz. With special circuitry, the same techniques can be used to measure frequencies up to 1 GHz.

The upper frequency-measuring limit of a digital counter is a function of the switching speed of the logic circuits, primarily the main gate and the input bcd counter. With special high-speed TTL logic circuits, counting frequencies up to about 50 MHz are possible. With ECL logic circuits, counting at a 1-GHz rate is possible. At these frequencies, the logic circuits as well as the input signal-conditioning circuits are critical. Special design and packaging techniques are required in order to achieve this counting rate.

The ability to measure very high frequencies gives the counter many applications in the communications field. For years, there has been a trend toward the use of higher radio frequencies, not only vhf and uhf but in the microwave region as well. Today, microwave equipment is common in many areas of communications. Radio transmitting and receiving equipment operating in the 400–500-MHz and 800–900-MHz ranges is very common. To design, service, and maintain such equipment properly, digital counters with a frequency-measuring capability in the range from 500 MHz to 1 GHz are required.

As more sophisticated rf equipment is developed, the need to measure even higher frequencies is created. Today, radio equipment and other communications gear can operate at frequencies up to 50 GHz. Counters have been developed to measure these frequencies. These counters use special techniques to achieve this frequency-measuring capability.

The purpose of this chapter is to introduce the various techniques used to extend the high-frequency measurement capability of a counter. A number of popular techniques are widely used. These are the prescaling, heterodyne, and transfer-oscillator methods. Each technique offers various advantages and disadvantages depending on the specific application.

This chapter also discusses some of the special problems involved in accurately measuring very low frequencies (lower than 1 Hz). Getting the necessary resolution requires some unique techniques. Several popular low-frequency measurement methods are described.

## PRESCALING

All of the techniques for measuring high frequencies involve a process that converts the high frequency into a proportional lower frequency that can then be measured with conventional counting circuitry. This translation of the high frequency into the lower frequency is called *down conversion*.

Prescaling is a down-conversion technique that simply involves the division of the input frequency by a factor which puts the resulting signal into the normal frequency range of the counter. For example, suppose that the maximum counting capability of the main gate and input bcd counter is 100 MHz. To measure a 400-MHz signal, the counter would have to be preceded by a frequency-divider circuit that generates an output frequency lower than 100 MHz. The resulting signal could then be counted. A 4-to-1 prescaler, or frequency divider, would accomplish this result. In practice, any integer division ratio can be used in a prescaler. But since most frequency-divider circuits are flip-flops which count in binary, most prescalers feature division ratios that are some power of 2, such as 2, 4, 8, 16, etc. Special high-frequency IC bcd counters which work as decade, or ten-to-one, frequency dividers are also widely available. By using a 10-to-1 prescaler on a 100-MHz counter, frequencies up to 1 GHz (1000 MHz) could be measured. Typically, prescaling techniques can be used successfully up to approximately 2 GHz with currently existing circuit technology.

Dividing the input frequency by some integer N will cause the counter to display a frequency that is lower by a factor of N. The contents of the bcd counters must then be multiplied by N in order for the display to read the correct input frequency. If a scaling factor of 10 is used, this multiplication process is easy. In effect, all you have to do is move the decimal point one position to the right in order to correct for the division by 10.

For example, assume that a 220-MHz signal is being measured. A 10-to-1 prescaler is used. The prescaler output is 220 ÷ 10, or 22 MHz, and that is what the counter display will read. This can be mentally multiplied by 10 to get the correct reading. In some commercial counters which incorporate a built-in 10-to-1 prescaling feature, the display correction is made by simply selecting the proper decimal-point indicator when the preselector is switched in.

If division factors other than 10 or some multiple of 10 are used, the correction of the display is more difficult. This can be accomplished by dividing the time base by the same integer factor used in the prescaler. The result will be a longer gate time that will compensate for the prescaling. The gate time will be extended by the same factor as the scaling factor so that the counter will show the correct frequency. Fig. 4-1 shows how this can be done. In this example, a ÷ 4 prescaler is used. The input frequency, f, is reduced to f/4 by the prescaler. Without some correction, a 220-MHz input would be displayed as 220/4 = 55 MHz. However, to make the display read correctly, the time base is divided by the same scaling factor prior to being applied to the control circuits. This generates a gating pulse four times the normal duration. A 100-millisecond gate would be extended to 400 milliseconds with this arrangement. Since the main gate is enabled four times longer, four

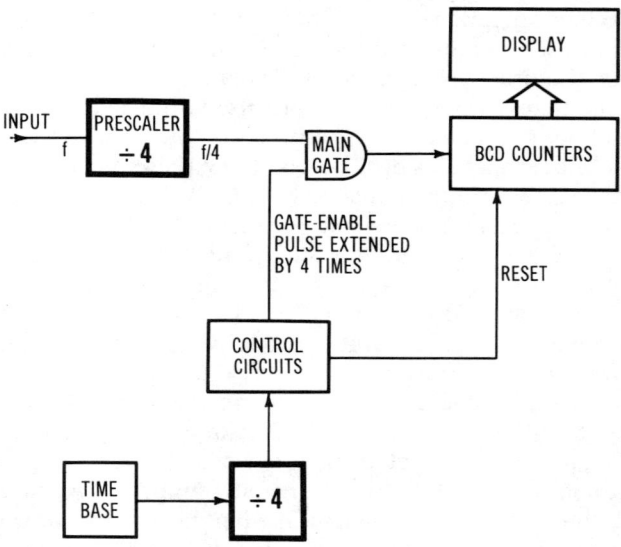

Fig. 4-1. Method of display correction used with prescaling.

times the normal number of input pulses are counted, and the display shows $4 \times 55 = 220$.

Most high-frequency counters that use prescaling have scaling factors of 10 or some multiple of 10. Even higher frequencies can be measured with prescalers of $\div 100$ and $\div 1000$.

It is important to realize that while prescaling does permit the measurement of higher frequencies, it is not without its disadvantages. One disadvantage is that resolution is lost. One digit of resolution is lost for each decade of prescaling incorporated. For example, assume that it is desirable to measure a 450,700-kHz (450.7-MHz) signal with a 100-MHz counter. This can be done by using a 10-to-1 prescaler that will bring the input frequency down within the range of the counter. If the input signal is passed through a 10-to-1 frequency divider, the counter will see an input frequency of 45.07 MHz. Given that each counter has a fixed number of decimal readout-display digits, and moving the decimal point to correct for the input scaling, you can see that one digit of resolution is lost. Assume the above-mentioned counter has 8 digits and the display reads 45070.000. To correct for the prescaling, the decimal point is moved one place to the right, or 450700.00. One digit of resolution is lost.

One way to overcome the loss of resolution in such a situation is to increase the gate time by the prescaling factor as illustrated earlier. If the counter you are using has a time base with switchable decade increments, you can simply select the next higher decade and multiply the gate time by 10. This adds back the single digit of resolution lost in the prescaling operation.

As you can see, achieving the same resolution of measurement with prescaling requires a longer measurement time. When time-base increments of less than 1 second are used, the operator does not typically notice the increase in measurement time. However, consider the measurement speed when a 1-second gate pulse is used. In order to achieve the same resolution obtained with a 1-second gate pulse and a factor of 10 prescaling, a 10-second gate time must be used. While 10 seconds may not appear to be a long period of time, when making electronic measurements it is significant. While the long measurement time is undesirable, this trade-off of measurement speed versus resolution may be necessary in order to achieve the desired resolution and accuracy of measurement.

The prescaling technique for extending the frequency-measuring capability of a counter is widely used. It is simple to implement with the modern high-speed integrated circuits available. It is also the most economical method of extending the counting range. Prescalers can be built into the counter and switched in when necessary. Alternatively, external prescalers which are widely available for low-cost counters can be used. Most prescalers operate in the range from 200 MHz to 2 GHz.

For frequencies beyond 2 GHz, other, more sophisticated down-conversion techniques must be used. Fig. 4-2 shows an older counter with prescaler plug-in accessories.

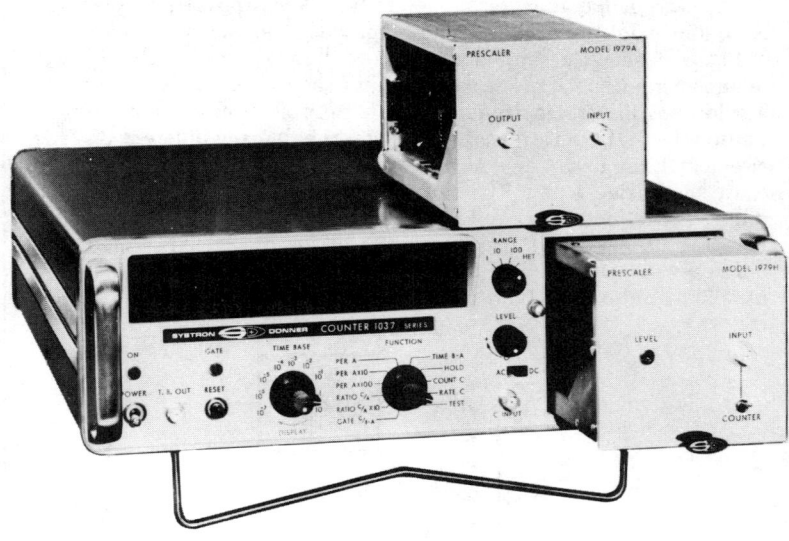

Courtesy Systron-Donner Corp.

Fig. 4-2. Digital counter with plug-in prescaler units.

## HETERODYNE DOWN CONVERSION

One of the earliest techniques used to extend the frequency range of a digital counter was the heterodyne conversion method. This approach uses standard mixing techniques common to most superheterodyne receivers. An internally generated local-oscillator signal is applied to a mixer along with the unknown input signal. The mixer generates the sum and difference of these two input frequencies. The difference frequency, of course, is lower than the frequency of the input signal. This lower-frequency signal can be applied to a digital counter for direct measurement. If the exact frequency of the local oscillator in the heterodyne converter is known, the frequency of the input signal can be calculated accurately from the value displayed by the counter. See Reference X for more information on mixing.

The heterodyne down-conversion technique depends on mixing to generate a lower-frequency signal that can be measured by standard digital counting circuits. Counters using this method have a front-end mixer ahead of the counter. In such counters, the frequency of the local

# REFERENCE X

## How Mixers Work

The basis of operation for the heterodyne down-conversion technique is nonlinear mixing. Mixing is an amplitude-modulation technique.

Fig. X-1 shows a simplified block diagram of a mixer circuit. The unknown input signal whose frequency is to be measured is designated $f_x$. The local-oscillator signal is designated $f_o$. Both of these signals are applied to the mixer. The mixer itself can be any one of many different types of circuits. It employs a nonlinear device such as a transistor or diode to cause the mixing action. The output of the mixer includes the sum and difference frequencies as well as the original input signals. A filter circuit, low-pass or bandpass, usually rejects all of the signals except the difference frequency. The difference frequency is the output that is usually amplified and then processed further in some way.

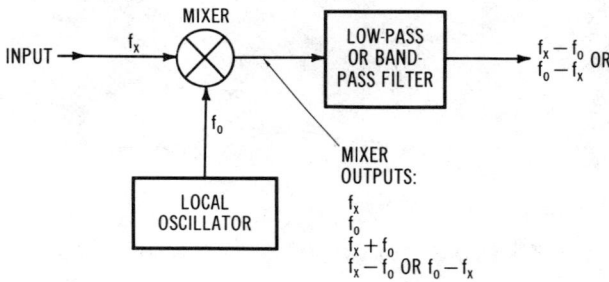

Fig. X-1. Simplified block diagram of a mixer.

oscillator is made variable so that a wide range of input frequencies can be accommodated. In this way, the mixer may be adjusted so that its output signal falls within the normal range of a conventional direct-counting counter. Assume that the local oscillator frequency, $f_o$, is less than the input frequency to be measured, $f_x$. The filter at the mixer output will select the difference frequency, $f_d = f_x - f_o$. The counter display will show this frequency. In order to determine the exact frequency of the input, the local-oscillator frequency must be added to the counter display; $f_x = f_d + f_o$. An example will best illustrate this. Assume that the input frequency to be measured is 780 MHz. The local-oscillator frequency is 700 MHz. The mixer output then is 780 − 700, or 80 MHz. For this example, the counter is capable of measuring up to 100 MHz. The 80-MHz output from the mixer is well within the range of the counter. The counter, of course, displays the frequency 80

MHz. To obtain the correct input frequency, the 700-MHz local-oscillator frequency must be added to the counter display.

The difficulty in implementing this technique is that to obtain an accurate measurement, the frequency of the local oscillator must be precisely known. It is extremely difficult just to generate a frequency in this range, much less provide high accuracy in frequency determination. It is not possible to calibrate a dial for a variable-frequency oscillator that will provide accuracy of frequency indication comparable to that obtainable with a frequency counter. For that reason, any measurement made with heterodyne down conversion is only as accurate as it is possible to determine the frequency of the local oscillator.

This problem can be overcome with special techniques. In modern counters that use heterodyne down conversion, an accurate local-oscillator frequency is obtained by using the counter time base and multiplying its frequency upward. The local-oscillator frequency applied to the mixer will be just as accurately known as the frequency of the time base.

There are two basic types of heterodyne frequency-conversion systems used with counters. The first type is a manual system in which the local-oscillator frequency is selected manually with a tuning dial of some sort. This manual selection method was used in early microwave counters. Often, the counter, a direct conversion unit, was manufactured to accept plug-in units. One of the plug-in units was a heterodyne converter capable of selecting and measuring high-frequency signals. A tuning dial was provided for frequency adjustment.

Modern heterodyne converters are automated, and no manual adjustments are necessary. Through the use of digital logic or microprocessors, the entire measurement technique is automated, thus providing fast, convenient measurement of very high frequencies with a sophisticated technique. Both the manual and automatic measurement systems are described in the following paragraphs.

## Manual Heterodyne Conversion

Fig. 4-3 shows a simplified block diagram of a manual harmonic converter. The 10-MHz crystal time base from the counter is applied to a frequency-multiplier circuit. The multiplier multiplies the time-base signal by 20, producing a 200-MHz signal. This signal is applied to a harmonic generator. The harmonic generator, usually a step recovery diode (see Reference XI), generates a wide range of harmonics of the 200-MHz input. The output of the harmonic generator is a "comb" of frequencies spaced at 200-MHz intervals. Fig. 4-4 shows what the crt output of a spectrum analyzer would show if it were monitoring the output of the harmonic generator. The harmonics generated could be as high as 12 GHz. This means that the comb generator produces signals from 200 MHz to 12 GHz in 200-MHz increments.

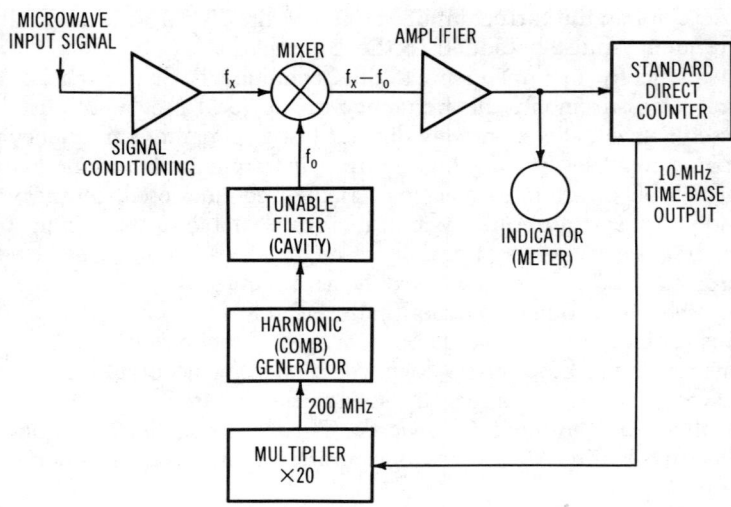

Fig. 4-3. Simplified block diagram of a manually operated heterodyne down converter.

The output of the comb generator is applied to a tunable filter. The purpose of the filter is to select one of the harmonic-generator output signals. This tunable filter is usually a manually adjustable microwave cavity. A tuned cavity is the microwave equivalent of a tuned inductor-capacitor circuit. The selectivity of the tunable filter is such that it can easily discriminate between the output signals from the harmonic generator spaced every 200 MHz. A calibrated tuning dial is provided to indicate which single harmonic is selected.

---

### REFERENCE XI

#### Step Recovery Diodes

A step recovery diode (SRD), or snap varactor, is a special pn junction diode designed for frequency-multiplication and comb harmonic-generation applications. When the SRD is forward biased, current passes through it as in a normal semiconductor diode. During this forward-biased state, current carriers are stored in the unique pn junction arrangement. When the SRD becomes reverse biased, current continues essentially unimpeded because of the stored carriers. But eventually the stored carriers are depleted. When this happens, the SRD rapidly "snaps" off to its high-impedance, or nonconducting, state. This turn-off step is very abrupt and is rich in high-order harmonics.

*Continued on next page.*

## REFERENCE XI (cont.)

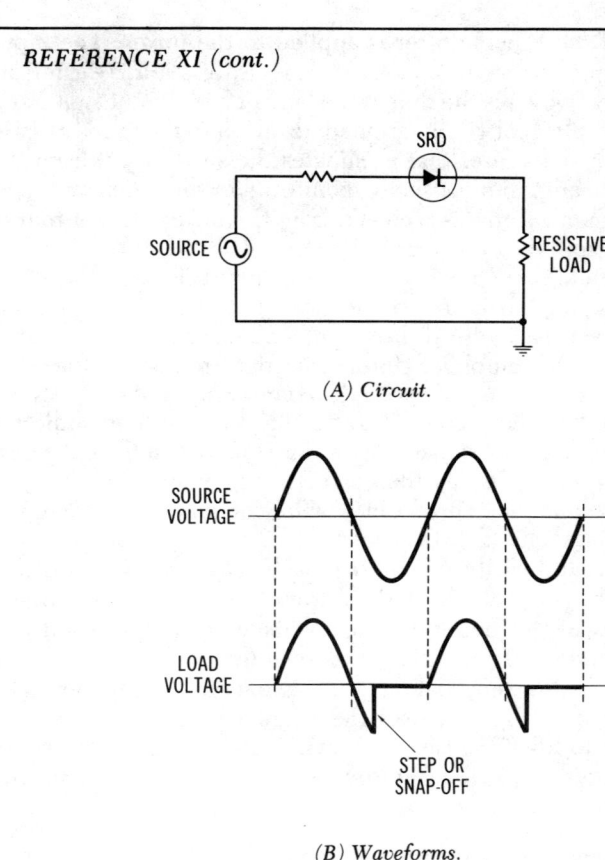

(A) Circuit.

(B) Waveforms.

Fig. XI-1. Basic SRD comb generator circuit.

Fig. XI-1 shows an SRD comb generator with the appropriate input and output waveforms. For frequency-multiplier applications, the resistive load is replaced by a tuned circuit or cavity.

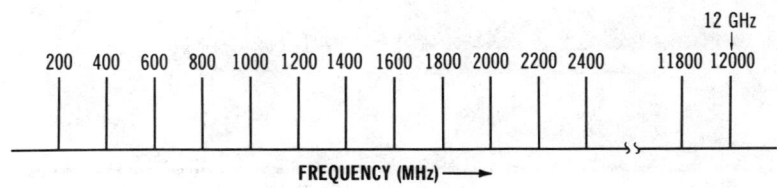

Fig. 4-4. Comb output of harmonic generator.

The output of the tunable filter is applied to the mixer. The input signal whose frequency is to be measured, $f_x$, is also applied to the mixer. The mixer output, which is the difference between the input signal and the local-oscillator signal, $f_o$, is applied to an amplifier, which also incorporates a low-pass filter that eliminates the sum output from the mixer $(f_x + f_o)$. The amplifier is usually monitored by an indicator circuit that will show when a signal is received from the mixer. The output of the amplifier is applied to a standard counter.

The measurement procedure is to apply the input signal to the mixer and then tune the tunable filter until the meter indicates a mixer output. As the cavity selects each one of the harmonic signals produced by the comb generator, the amplifier filter will determine whether the difference signal is within the range of the counter input. As soon as the proper harmonic has been selected by tuning, the amplifier indicator will show the presence of a signal. It is at this time that sufficient signal reaches the counter to display the correct frequency.

The frequency being measured can now be computed by adding the frequency of the harmonic $(f_o)$ to the counter display. Usually the harmonic frequency is indicated on the tuning dial of the cavity. This is a highly accurate frequency since it was derived from the 10-MHz time base.

Early versions of the heterodyne conversion technique used this manual tuning system. In addition, this technique requires not only manual tuning but also computation of the actual input frequency. The counter illustrated in Fig. 4-5 uses the manual heterodyne method. Frequencies up to 18 GHz can be measured with this counter. In modern counters that use the heterodyne conversion technique, the tuning and computation functions have been automated.

### Automatic Heterodyne Conversion System

An automated version of the heterodyne conversion technique is shown in Fig. 4-6. It is similar to the manual system described before but with several changes and additions.

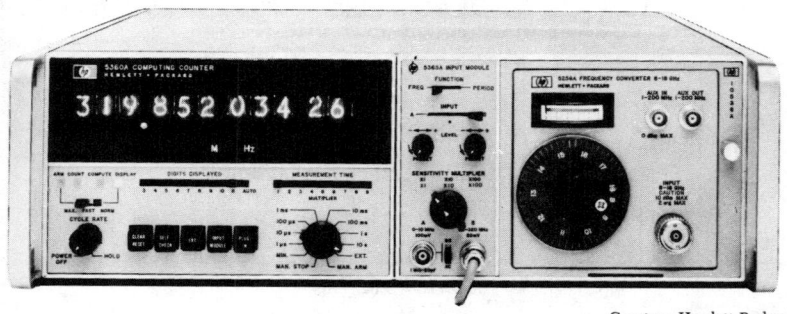

Courtesy Hewlett-Packard

Fig. 4-5. Counter using the manual heterodyne down-conversion method.

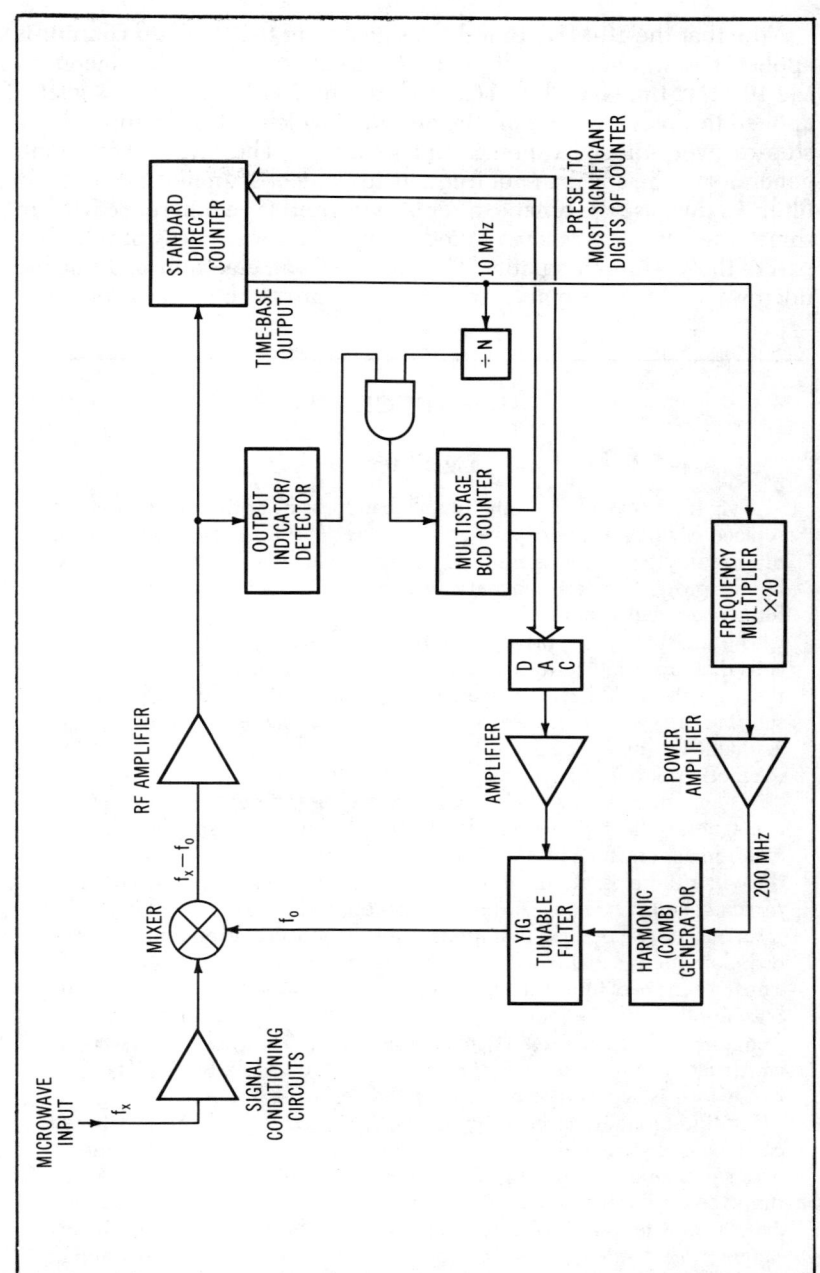

Fig. 4-6. A microwave counter using automatic heterodyne down conversion.

Note that the 10-MHz time-base signal from the standard counter is applied to a frequency multiplier. The frequency multiplier increases the 10-MHz time base by a factor of 20. The 200-MHz output signal is applied to a power amplifier, the output of which in turn is applied to a step-recovery-diode harmonic comb generator. The output of the comb generator is the spectrum of frequencies which is applied to a tunable filter. In this case the tunable filter is a yig filter. (See Reference XII.) In some cases the step-recovery-diode harmonic generator is physically a part of the yig filter assembly. The yig filter is an electronically tunable microwave device capable of selecting one of the comb-generator

---

### REFERENCE XII

### Yig Filters

A yig filter is an electrically tunable bandpass filter that can be used to replace cavities in microwave applications. The main element in a yig filter is an yttrium-iron-garnet (yig) crystal. This is a ferrite material that has the property of ferromagnetic resonance. The main element of the yig filter is a small sphere of the yig material.

The yig sphere is typically mounted in a housing which also contains coils that form an electromagnet which produces a magnetic field around the yig sphere. When the strength of the magnetic field is adjusted to saturate the yig sphere, the sphere acts as a classical resonant circuit. The resonant frequency of the sphere is controlled by the intensity of the magnetic field applied to it.

When a microwave signal is applied to the yig sphere, the rf field sets up a magnetic effect perpendicular to the dc bias field supplied by the electromagnet. If the bias field is adjusted so that the natural resonance of the yig sphere is at the input frequency, a strong interaction called *ferromagnetic resonance* occurs. At this time, the input is readily passed to an output pickup loop mounted in the yig housing. Frequencies above or below the resonance point of the yig sphere are greatly attenuated. Thus, a bandpass filter is formed. To vary the resonant frequency, the dc bias applied to the electromagnets is changed. The filter resonant frequency is linearly proportional to the dc bias. Typical yig filters can be constructed to operate over the range from 400 MHz to 40 GHz. The bandwidth is typically 0.2 to 2% of the resonant frequency.

Yig filters offer three main advantages over mechanically tuned cavities. First, yig filters can be significantly smaller. Second, they offer a wider dynamic range of signal-handling capability. Third, they can be tuned electronically much more easily and, therefore, permit automation of the tuning function. Yig filters can also be tuned over a wider frequency range. A mechanically tuned cavity is usually limited to a 2-octave tuning range. On the other hand, yig filters can be tuned over a 5-octave range.

harmonics. The yig filter is essentially a bandpass filter that can be adjusted by the application of external dc voltage. This allows the tuning process to be automated. Some microwave counters use hybrid resonator filters switched in and out by pin diodes.

The output of the yig filter is, of course, applied to the mixer along with the input signal. The output difference signal $(f_x - f_o)$ from the mixer is applied to an rf amplifier, which drives an output indicator/detector circuit. The output of this indicator circuit is a dc voltage used to control the tuning of the yig filter. The indicator/detector circuit output is used to enable an AND gate, which controls the application of clock pulses derived from the time base through a divider. When no signal is present at the mixer output, the AND gate is enabled.

The control signal for tuning the yig filter is derived from a bcd counter and a digital-to-analog converter (dac). As the bcd counter is incremented, it generates a binary number, which is applied to the dac. The output of the dac is a staircase voltage, which is amplified and used to drive the yig filter.

The measurement process begins when the microwave input signal is applied to the mixer. The bcd counter circuit is initialized, and the output of the dac is a dc voltage which causes the yig filter to select its lowest harmonic. With this harmonic input to the mixer, a specific mixer output signal will be generated. If it is not within the bandwidth of the rf amplifier, the indicator/detector signal will be such that it will enable the AND gate and cause the bcd counter to begin incrementing. As the counter is stepped, an increment of voltage will be generated at the output of the dac, thus tuning the yig filter and selecting the next higher harmonic from the comb generator. The counter will continue to count and increase the voltage and select each of the harmonic frequencies in sequence. Whenever the correct harmonic is selected the rf-amplifier indicator will show that a signal within the frequency range of the amplifier has been selected. At this point, the indicator/detector output inhibits the AND gate, and the counter is stopped. The output of the rf amplifier is measured by conventional counting methods, and the frequency is displayed.

Keep in mind that in order for the counter display to read correctly, the difference output from the mixer, $f_d = f_x - f_o$, must be added to the yig-filter output, $f_o$. This can be accomplished in a variety of ways. Here, the bcd output signal from the counter is used to preset the most significant digits on the counter display. This in effect causes the correct harmonic frequency, $f_o$, to be added to the difference frequency from the rf amplifier so that the correct measured frequency is displayed. The reason for this is that the value of the number in the bcd counter driving the dac yig-filter control circuit is set up to be the same as the harmonic selected by the filter. For example, assume an input frequency $f_x =$ 12.763 GHz. Assume further that when the bcd counter is incremented to 12, the 12-GHz harmonic is selected. The mixer output is $f_x - f_o$, or 12.763 − 12 = 0.763 GHz, or 763 MHz. The conventional counter has eight digits and displays 763000. The most significant digits are preset to the bcd counter contents of 12 for a final display of 12,763,000 kHz, or 12,763.000 MHz. With this arrangement, not only is the tuning function carried out automatically, but also the correct frequency of the input signal is displayed directly, and no further computation by the operator is required.

Remember that the configuration shown here is only one of many possible arrangements. In modern counters, a microprocessor can replace a lot of the bcd counting, control, and final frequency-computing functions.

## TRANSFER-OSCILLATOR DOWN CONVERSION

Another technique widely used for converting a high-frequency signal into a signal compatible with a frequency counter is the transfer-oscillator (TO) method. This down-conversion technique, like the heterodyne method, also uses a mixer circuit. The unknown frequency, $f_x$, is applied to the mixer along with the output from a variable-frequency oscillator (vfo) as shown in Fig. 4-7. The vfo output frequency, $f_o$, is multiplied by a frequency multiplier before it is applied to the mixer. The idea is to generate a frequency, $Nf_o$, equal to the frequency of the unknown signal. For example, if the frequency of a 5-GHz signal is to be measured, the vfo and the multiplier should be capable of generating a 5-GHz input signal to the mixer. The output of the mixer is monitored by a detector, either an oscilloscope or earphones.

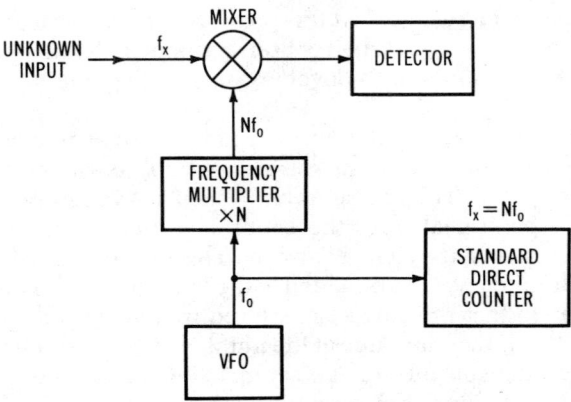

Fig. 4-7. Block diagram of transfer-oscillator high-frequency measurement technique.

Remember that the mixer generates the sum and difference frequencies. If the vfo-generated signal is equal to the input signal, then the difference frequency will be zero. By tuning the vfo, it is possible to create a difference output in the audio-frequency range which can be heard in earphones. To make the measurement of the input signal, the vfo is tuned for a zero-beat condition, where $f_x = Nf_o$. As the vfo frequency becomes closer and closer to the incoming frequency, the difference frequency becomes smaller and the audio tone lower. When the vfo signal frequency is equal to the incoming signal frequency, the difference frequency will be zero, and no signal will be heard in the earphones. Since most individuals do not hear frequencies below approximately 30 Hz, some error in measurement is possible. That is why it is desirable to use an oscilloscope to monitor the mixer output so that the exact zero-beat condition can be noted.

When the zero-beat condition exists, the input signal frequency generated by the vfo and the multiplier is exactly equal to the incoming signal frequency. The frequency of the vfo, $f_o$, which is within the range of a normal counter, can be used as a measure of the input frequency. By multiplying the counter reading ($f_o$) by the multiplication factor (N), the exact frequency of the input signal is obtained ($f_x = Nf_o$).

As an example, assume that the input frequency is 5 GHz. Also assume that the vfo tunes over the range 50 to 150 MHz. Further assume that the multiplier increases the vfo output by a factor of 50. As the vfo is tuned, an audible signal will be heard when the vfo frequency approaches 100 MHz. When the vfo is exactly at 100 MHz, the zero-beat condition will be detected. At this point, the conventional counter will read 100 MHz. Multiplying this by multiplier factor 50 gives the input frequency of $50 \times 100 = 5000$ MHz, or 5 GHz.

This method of measuring high-frequency signals is called the transfer-oscillator method simply because the technique causes the input frequency to be transferred indirectly to the vfo. The lower-frequency vfo can then be adequately measured by more conventional counting methods. This technique has been implemented in a variety of ways in digital counters. Manual transfer-oscillator plug-in units were widely used in earlier uhf counters. The manual transfer-oscillator plug-in units effectively implemented the basic transfer-oscillator circuit shown in Fig. 4-7. However, most modern uhf counters incorporate a fully automatic transfer-oscillator down-conversion circuit. We will take a look at both the manual and automatic versions of this widely used down-conversion method.

**Manual Transfer Oscillator**

Most manual transfer-oscillator conversion units were made as plug-ins for standard lower-frequency counters. The term "manual" refers to the fact that an operator must tune the vfo himself in order to determine the zero-beat condition. Once this has been done, the main counter provides the correct output frequency.

Most of the transfer-oscillator plug-ins provide additional features that simplify and speed up the measurement process. They also make it more accurate. One of the most critical components of the transfer-oscillator unit is the vfo. An extremely stable tunable oscillator must be used if accurate measurements are to be made. Most tunable oscillators are made with inductive and capacitive networks whose stability does not compare with the stability of the crystal time base in the counter. In many cases, the vfo frequency will change over a short period of time even after a zero-beat condition has been detected. The frequency may change before an accurate measurement can be obtained. The modern transfer-oscillator plug-in unit overcomes this stability problem. In addition, the transfer-oscillator plug-in also provides for automatic multiplication of the counter reading to provide an accurate, exact reading of the frequency on the display.

Fig. 4-8 shows a block diagram of a typical manual transfer-oscillator unit. In this circuit, the vfo is replaced by a voltage-controlled oscillator (vco). This vco is both manually tunable and controllable by a dc input. The vco drives a harmonic generator which feeds a comb signal to the mixer. The unknown signal is also applied to the mixer. The mixer output, which is the frequency difference between the two signals, is applied to a video amplifier. The output of this amplifier is provided so that an external oscilloscope can be used to detect the zero-beat or minimum-difference condition.

Notice that the video-amplifier output is also applied to a phase detector. A stable time-base output signal from the digital counter is applied to the other phase-detector input. A low-pass filter (lpf) filters the output of the phase detector into a dc voltage that controls the vco. The resulting circuit is a phase-locked loop (pll). (See Reference XIII for details on pll's.) The only difference between this circuit and a conventional pll is that the phase-detector input from the video amplifier replaces the vco output, which is usually the feedback source. Here the feedback occurs through the mixer.

The purpose of this arrangement is to provide high stability of the vco. The stability of the vco in a phase-locked-loop circuit will be equivalent of the stability of the input reference signal, which in this case comes from the counter time base. This arrangement overcomes the vfo/vco stability problem mentioned earlier.

Note also in Fig. 4-8 that it is possible to disable the phase-locked-loop circuit by disconnecting the vco from the phase detector and lpf. This allows the vco to be tuned manually without external electrical control.

The output of the vco, which is usually in the range from 100 to 200 MHz, is applied to a harmonic generator. The harmonic generator is a comb-generator multiplier made with a step recovery diode. The

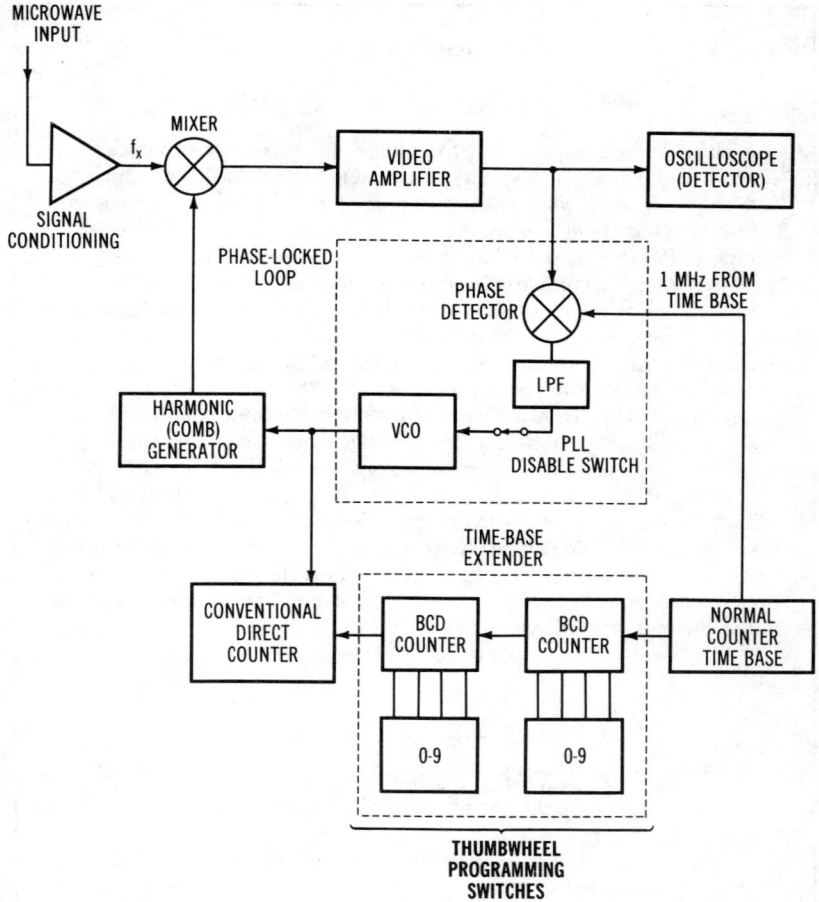

Fig. 4-8. Manual transfer-oscillator circuit for measuring microwave frequencies.

harmonic generator can produce harmonics as high as the 100th harmonic. The comb-generator output is applied to the mixer along with the unknown input signal, $f_x$.

Another feature of the manual transfer-oscillator conversion unit is an automatic time-base extender feature. This prevents the operator from having to multiply the vco frequency by the harmonic in order to obtain the correct frequency. The time-base extender simply extends the gate time of the counter by a factor equal to the harmonic. This results in the vfo frequency being counted for the correct number of harmonic intervals; thus, the counter reading is automatically multiplied by the harmonic, and the correct input frequency is produced on the display.

# REFERENCE XIII

## Phase-Locked Loop

A phase-locked loop (pll) is a frequency- and phase-sensitive feedback control circuit used in a variety of frequency-synthesis, demodulation, and frequency-multiplication applications. Fig. XIII-1 shows the basic pll circuit. It consists of a phase/frequency sensitive mixer or detector, a low-pass filter (lpf), and a voltage-controlled oscillator (vco). In operation, the input signal is compared in the mixer to the vco output, which is at approximately the same frequency. If there is a phase or frequency difference between the two, the mixer generates an error output signal. The lpf filters this into a dc voltage that is used to control the vco. The dc control voltage for the vco is proportional to the amount of error at the mixer output. The error, in turn, is proportional to the amount of phase or frequency difference between the input signal and the vco output. The pll is designed so that the dc control voltage forces the vco frequency to change in a direction that minimizes the error. The vco output will vary until it is equal in frequency to the input. When this condition occurs, the two signals are said to be synchronized, or "locked." If the input frequency changes, the error increases, and the dc control voltage causes the vco frequency to change so that it equals the input frequency. As the input changes, the pll vco output "tracks" it.

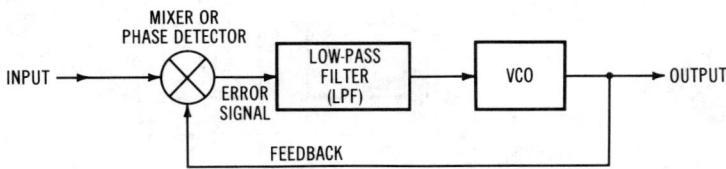

Fig. XIII-1. Basic phase-locked-loop circuit.

A popular application for the pll is a frequency multiplier that increases an input frequency $f_i$ by some integer N. The vco is set to the desired higher output frequency, $f_o$, which is N times the input, or $f_o = Nf_i$. The vco output is then divided by N before it is applied to the mixer, as shown in Fig. XIII-2. With this arrangement, the pll produces an output frequency N times the input frequency. If the input frequency changes, the multiplied output changes because of the tracking characteristics of the pll.

The circuit configuration shown in Fig. XIII-2 can also be used as a frequency synthesizer. A synthesizer is a signal generator whose output can be varied in discrete increments over a wide range. In frequency-

*Continued on next page.*

*REFERENCE XIII (cont.)*

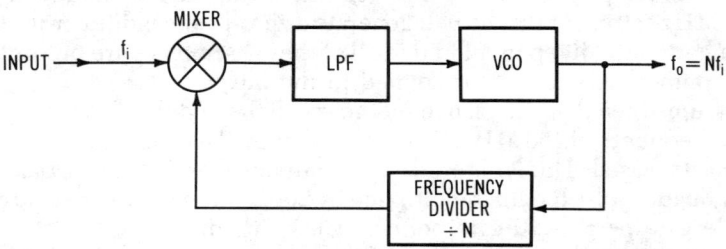

Fig. XIII-2. Phase-locked loop connected for multiplier or synthesizer use.

synthesizer applications, the pll input is derived from a precision crystal oscillator to obtain highly accurate output frequencies. To vary the output frequency in steps, the frequency divider in the feedback loop is made variable in increments. Flip-flops and counters when used as dividers can be made switchable so they can be set to divide by any integer N. If thumbwheel switches and bcd counters are used in the divider, a frequency-selectable synthesizer is formed.

The time-base extender is simply a programmable frequency divider inserted between the time-base oscillator and the counter control circuits which generate the gate-enable pulse. This frequency divider can be set to divide by any factor between 1 and 100. This can usually be accomplished by two bcd counters used as frequency dividers. The division ratio can be preset by external thumbwheel selector switches. The harmonic must be manually determined and then entered on the thumbwheel switches to provide correct time-base extension.

The following procedure is used to measure an unknown frequency with the manual transfer-oscillator converter unit.

First, set the thumbwheel switches on the time-base extender to the harmonic to be used. This is easy to do, because the harmonic can be determined if the approximate input frequency is known. For example, assume that you wish to measure a 3-GHz signal. Knowing this, you can determine the harmonic, N, by knowing the frequency range of the vco. If the vco has a frequency range of 100 to 200 MHz, then you can compute the harmonic by dividing the unknown input frequency, $f_x$, by the vco frequency, $f_o$.

$$N = \frac{f_x}{f_o}$$

As you can see, there are a number of harmonics that can be used depending on the exact frequency setting of the vco. For example, if the vco frequency is 100 MHz, then a harmonic of 30 must be used to generate a 3-GHz input signal to the mixer to match the unknown input and create a zero-beat condition. On the other hand, a 200-MHz setting of the vfo can produce a zero-beat condition with the 15th harmonic. A 150-MHz setting of the vfo will generate a zero-beat condition with the 20th harmonic. Keep in mind that all of these harmonics are present in the comb-generator output applied to the mixer.

Assume then that you plan to set the vco to the middle of its range and use a frequency of 150 MHz. To get a frequency of 3 GHz, a multiplier of 20 must be used. This is entered on the thumbwheel switches so that the gate-enable pulse length is multiplied by 20. In other words, instead of a single gate period, 20 gate periods will be used.

Next, disable the phase-locked-loop circuit with the switch provided. Tune the vco manually until a zero-beat condition is noted on the oscilloscope. Once the zero-beat condition has been obtained, re-enable the phase-locked loop with the pll disable switch. At this time, the phase detector will compare the outputs from the video amplifier and the time-base circuit and will generate a dc voltage that will adjust the vco automatically and lock it on frequency.

It is important to note that a phase-locked condition cannot be obtained with one of the pll phase-detector inputs at zero. In fact, the phase-locked condition will occur only when the frequencies of the two phase-detector inputs are equal. For that reason, this circuit is designed to operate correctly when the difference output from the mixer and video amplifier is 1 MHz instead of zero. One input to the phase detector is the 1-MHz time-base signal. The other input to the phase detector is a 1-MHz difference signal from the mixer. The output voltage from the phase detector will be such that it will set the frequency of the vco to a value that will produce a difference frequency of 1 MHz with the unknown input. Under these conditions, the circuit is locked. Changing the vco frequency, of course, removes this balanced, or locked, condition. An error signal will be generated to vary the vco until a 1-MHz difference signal is generated.

In actual operation, the vco must initially be tuned by hand, with the pll disabled, to produce a 1-MHz output instead of zero beat. This 1-MHz output condition is detected in a variety of ways in various commercial counters. However, once this 1-MHz condition is noted, the pll is again enabled. The pll then locks, making the vco stability equivalent to that of the 1-MHz time-base input.

At this time, the frequency of the vco harmonic is 1 MHz lower than the actual input frequency. The vco frequency is measured on the frequency counter. The 1-MHz error is corrected by adding 1 MHz to the value in the main bcd counters in the counter unit. This can be accomplished by presetting 1 MHz into the decade counters instead of resetting them to zero prior to making the frequency measurement.

Keep in mind also that the time-base extender has extended the gating period of the counter to multiply the vfo signal by the correct harmonic automatically. The counter display at this time correctly reads the input frequency.

## Automatic Transfer-Oscillator Converter

The transfer-oscillator down-conversion technique can be fully automated, thereby eliminating the need for the operator to perform manual tuning and harmonic selection. Most modern microwave frequency counters are fully automatic, and many use the automatic transfer-oscillator technique. In a counter using the automatic transfer-oscillator method, a voltage-controlled oscillator is automatically swept in frequency until the phase-locked loop of which it is a part becomes locked to the input signal. It is the frequency of this vco that is measured by the standard counting portion of the unit. Another separate set of circuitry in the counter is used for automatic determination of the harmonic selected. Once the harmonic has been determined, the time-base extender is automatically adjusted to provide the correct count. The overall result is the display of the correctly measured frequency on the readout. Such automatic microwave counters are just as simple to use as standard direct frequency counters at the lower frequencies.

Fig. 4-9 shows a block diagram of one type of automatic transfer-oscillator down converter. The unknown input signal whose frequency, $f_x$, is to be measured is applied to a power divider. The power divider is simply a resistive network that maintains the correct impedance level of 50 ohms while splitting the input signal into two equal signals. The two output signals are then applied to two separate sampler circuits.

The sampler circuit is simply a diode network which performs both the functions of harmonic generation and mixing. The term "sampler" is derived from the fact that such circuits are used as sampler or signal-acquisition circuits in high-frequency sampling oscilloscopes. The sampler mixes the unknown input signal ($f_x$) with harmonics of the vco signal ($f_1$). The frequency of $vco_1$ is in the range from 100 to 200 MHz. The sampler generates harmonics that are mixed with the incoming signal. The output of the sampler, or mixer, is a difference frequency in the range from 1 to 20 MHz. This signal is amplified in a video-amplifier circuit and applied to one input of a phase-detector circuit. The other input to the phase-detector circuit is a 10-MHz time-base signal for the reference. In order for a phase-locked condition to occur, the input signals to the phase detector must be equal in frequency. This means that the difference between the unknown input frequency and the harmonic of the vco must be 10 MHz. Whenever the video-amplifier output is 10 MHz, a phase-locked condition will exist. The output of the phase detector is filtered into a dc voltage which controls $vco_1$.

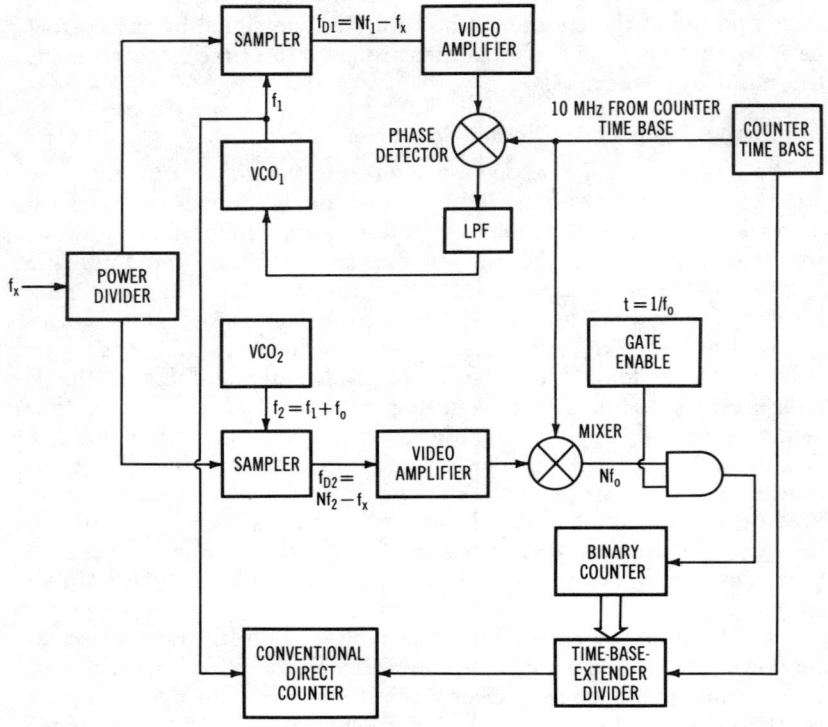

Fig. 4-9. Block diagram of an automatic transfer-oscillator down-conversion counter.

The relationship of the various frequencies in the phase-locked loop is expressed by the equation

$$Nf_1 - f_x = 10 \text{ MHz}$$

In this equation, N is the harmonic, or integer, by which the output of $vco_1$ ($f_1$) is multiplied. Rearranging the expression algebraically allows us to express the input frequency in terms of the vco frequency, its harmonic, and the time-base reference signal.

$$f_x = Nf_1 - 10 \text{ MHz}$$

The vco output, $f_1$, can be measured by direct counting techniques to indicate the frequency of $vco_1$, which is a function of the unknown input signal. However, the harmonic number is unknown, and, therefore, the correct input frequency cannot be fully determined. Also, correction must be made for the 10-MHz offset error. The remainder of the circuitry in Fig. 4-9 is used to determine the harmonic number, N.

The unknown input signal ($f_x$) is also applied to a second sampler.

Again, this sampler performs the harmonic-generation and mixing functions. The other input to this sampler/mixer is a second voltage-controlled oscillator, $vco_2$, whose frequency is nearly the same as the frequency of $vco_1$. The frequency of $vco_2$ is offset from $vco_1$ by a known frequency factor, $f_o$. The output frequency of $vco_2$ ($f_2$) is given by the expression

$$f_2 = f_1 + f_o$$

The difference frequency, $f_{D2}$, of the second sampler/mixer is given by the expression

$$f_{D2} = Nf_2 - f_x$$
$$f_{D2} = N(f_1 + f_o) - f_x$$

This signal is amplified in a video-amplifier circuit and is then mixed with the 10-MHz time-base reference signal in a separate mixer circuit. The difference output is the offset frequency, $f_o$, multiplied by the harmonic, N. The harmonic number, N, can now be determined by counting the output of this mixer circuit for a length of time equal to the period of the offset frequency, $f_o$. (Typical offset frequencies used in microwave frequency counters range from 1 kHz to 20 kHz.) The output of the mixer circuit is applied to a gate, which is enabled for a duration equal to the period of the offset frequency. The output of this gate is accumulated in a binary counter. The content of this counter then is a number, N, which is equal to the order of the harmonic. This number is transferred to the time-base-extender frequency-divider circuits. The time base is thus adjusted so that the main gate that counts $f_1$ is extended by N times so that the display reads correctly.

Keep in mind that the 10-MHz offset must be preset into the regular direct-counting bcd divider circuits in order for the correct frequency to be displayed.

The automatic transfer-oscillator frequency counter is one of the most complex and sophisticated instruments available today. It combines microwave technology and high-speed digital circuitry to make microwave frequency measurements as fast and easy as low-frequency measurements with a standard counter. Fig. 4-10 shows a commercial counter using a version of the transfer-oscillator down-conversion technique.

## THE HARMONIC-HETERODYNE CONVERTER

Another popular down-conversion technique used in microwave counters is the harmonic-heterodyne method. Fig. 4-11 shows a simplified block diagram of a harmonic-heterodyne converter unit. As with other down-conversion techniques, this unit is essentially a front end for a conventional counter. Like the harmonic-conversion

Courtesy Systron-Donner Corp.

Fig. 4-10. Microwave counter using automatic transfer-oscillator technique.

technique, this method converts the input signal into a lower frequency that can be measured by a conventional counter. However, other computations must be made in order to adjust the mixer output so that the correct frequency is displayed.

Referring to Fig. 4-11, you can see that a frequency synthesizer (Reference XIII) is used to generate a signal of known frequency $f_s$, which is applied to a sampler. The sampler generates a broad range of harmonics and mixes them with the unknown input signal of frequency $f_x$. The output of the mixer is a lower-frequency difference signal of frequency $f_D = f_x - Nf_s$, which is applied to a video amplifier. The output of the video amplifier is fed to a bandpass filter. At the output of the bandpass filter is a signal-detector circuit that indicates the presence or absence of a signal within the bandpass-filter range.

The frequency synthesizer is usually a phase-locked loop consisting of a highly stable vco whose frequency can be changed in discrete increments. Most frequency synthesizers are digital in nature and highly stable since the output frequency is referenced to an accurate crystal-oscillator reference such as the time base in the main counter.

The result of this arrangement is an output frequency, $f_D$, that is a function of the unknown input signal, $f_x$, the synthesizer frequency, $f_s$, and the specific harmonic, N, selected by the circuit. This is expressed by the equation

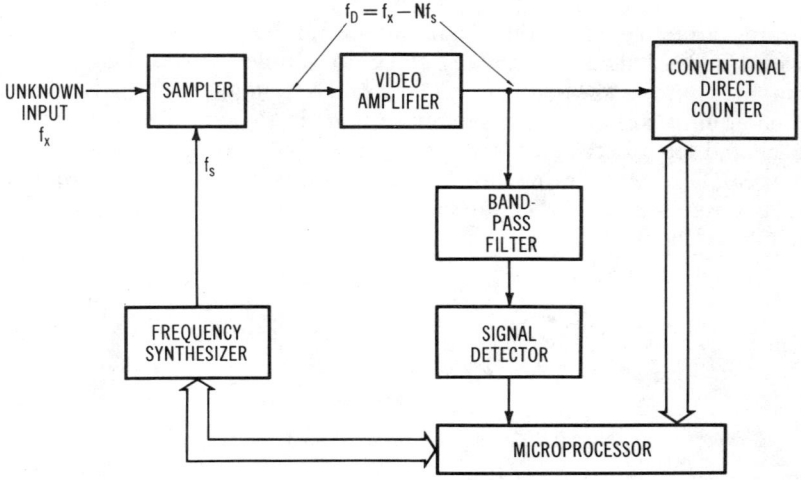

Fig. 4-11. Block diagram of the harmonic-heterodyne down-conversion technique.

$$f_D = f_x - Nf_s$$

Note in Fig. 4-11 that a block labeled "microprocessor" is included. This microprocessor provides the digital-logic control signals to operate the unit properly and perform the necessary computations, as you will see later. Note also that it is the output of the mixer and video amplifier that is fed to the input of the conventional counting circuit and measured. The output of the video amplifier, $f_D$, is not the correct frequency, but is related to it. The correct frequency is determined by further modifying the contents of the counter by the calculations made in the processor:

$$f_x = f_D + Nf_s$$

The operation of this unit begins when the microprocessor causes the synthesizer to be stepped, or incremented in frequency. The synthesizer scans across a predetermined frequency range. The synthesizer output is applied to the sampler, which generates a broad spectrum of harmonics. These harmonics mix in the sampler with the unknown input signal and generate a difference frequency, $f_D$. When the difference frequency is within the bandpass-filter range, the signal detector will indicate an output and thus signal the microprocessor to stop the scanning of the frequency synthesizer. The difference frequency is counted in the standard counter. The synthesizer frequency, $f_s$, and the difference frequency, $f_D$, are now known. It is only necessary to determine the harmonic number, $N$, in order to compute the unknown frequency, $f_x$.

The main problem in the harmonic-heterodyne converter then centers around determining the harmonic number, N. This can be done with a second signal source and separate sampler as was done in the automatic transfer-oscillator converter described earlier. Another technique is to have the processor automatically step the synthesizer back and forth between two closely spaced but known frequencies, $f_{s1}$ and $f_{s2}$. Then by observing the difference frequencies, $f_{D1}$ and $f_{D2}$, created in the reading on the counter, the processor can calculate N. The equations below show one way in which this might be accomplished.

$$f_{D1} = f_x - Nf_{s1} \quad \text{and} \quad f_{D2} = f_x - Nf_{s2}$$
$$f_x = f_{D1} + Nf_{s1} \quad \text{and} \quad f_x = f_{D2} + Nf_{s2}$$

Therefore:

$$f_{D1} + Nf_{s1} = f_{D2} + Nf_{s2}$$
$$f_{D1} - f_{D2} = Nf_{s2} - Nf_{s1}$$
$$f_{D1} - f_{D2} = N(f_{s2} - f_{s1})$$

$$\frac{f_{D1} - f_{D2}}{f_{s2} - f_{s1}} = N$$

Since $f_{s1}$, $f_{s2}$, $f_{D1}$, and $f_{D2}$ are known, N can be computed by the microprocessor.

Once the value of the harmonic has been determined, the unknown frequency can be computed. This is done by having the processor multiply the harmonic by the synthesizer frequency, $f_s$. The resulting number is then added to the difference frequency, $f_D$, which is contained within the counter. The result is the unknown frequency, $f_x$, which is then displayed.

The main advantage of the harmonic-heterodyne converter technique is that it can be made less expensive, since only one microwave component, the sampler, is needed. The precision microwave components are the most expensive part of any high-frequency counter. In addition, the necessary calculations can be carried out by a readily available low-cost microprocessor. The processor provides additional benefits by making the unit entirely automatic. A commercial counter using this technique is shown in Fig. 4-12.

## SPECIFICATIONS OF HIGH-FREQUENCY COUNTERS

The specifications of high-frequency or microwave counters are essentially the same as those of more conventional direct counters. However, there are some differences. Besides measuring a wider range of frequencies and typically having more digits of readout, high-frequency counters also differ in the way their specifications are expressed. In addition, there are several new specifications for high-frequency

Courtesy Hewlett-Packard

**Fig. 4-12. A microwave counter using the harmonic-heterodyne down-conversion method.**

counters that are not applicable to conventional lower-frequency counters. These include such things as measurement speed, fm and am tolerance, and amplitude discrimination.

In this section we take a look at those specifications particularly applicable to high-frequency counters. Then the various advantages and disadvantages of the various types of down-conversion techniques are summarized.

### Frequency Range

The upper frequency limit of microwave counters ranges from approximately 1 GHz to 40 GHz. Direct counters are capable of up to 500 MHz to 1 GHz. Prescaling techniques are valid up to approximately the 1–2-GHz range. The heterodyne converter can handle signals up to approximately 20 GHz. The transfer-oscillator and harmonic-heterodyne converter techniques are capable of covering signals up to 40 GHz.

### Number of Digits

Since it is desirable to measure microwave-frequency signals with good resolution, it follows that microwave counters will have more display digits than conventional lower-frequency counters. Most microwave counters have from 9 to 15 display digits that allow frequencies this high to be measured with excellent accuracy and resolution.

## Input Impedance

Because of the high frequencies involved, high input impedances are not practical in microwave counters. Even the smallest amount of stray, distributed, or shunt capacitance will greatly attenuate a microwave signal. Therefore, input impedances must be very low to minimize this effect. Almost universally, microwave counters use a standard 50-ohm input impedance.

## Sensitivity

As in conventional lower-frequency counters, sensitivity refers to the minimum level of the input signal that can be reliably measured. Unlike low-frequency counters whose sensitivity is measured in millivolts, high-frequency counters typically have sensitivity expressed in terms of dBm (Reference XIV). The typical sensitivity range for microwave counters is $-10$ to $-40$ dBm.

---

### REFERENCE XIV

### dBm

The term "dBm" is a way of expressing the measurement of power. Specifically, dBm is a power level in decibels referenced to a level of one milliwatt (mW).

$$dBm = 10 \log \frac{P}{1 \text{ mW}}$$

For example, 0 dBm is equal to 1 mW. Usually, the definition is valid only for measurements of power in a 50-ohm load.

---

## Dynamic Range

The dynamic range is a measure of the difference between the highest and lowest sensitivity levels of the counter. The highest sensitivity level essentially refers to the maximum input signal which can be tolerated by the counter and measured reliably. The dynamic range is ordinarily expressed in decibels. Typical dynamic ranges run from approximately 30 to 50 dB.

## Accuracy

The accuracy of a microwave counter is determined primarily by the same factors that affect conventional counters. The $\pm 1$ count error is a factor in microwave counters as well, but it is less important. It is the time-base accuracy that essentially determines the accuracy of a microwave counter. Highly stable time bases are desirable for high-accuracy measurements.

## Measurement Speed

Measurement speed is a specification unique to microwave counters. In addition to the time ordinarily required to measure a signal in a standard counter, a microwave counter has an additional *acquisition period*. To measure a frequency in any counter, a gate signal must be generated and input pulses passed to the decade counters. The same is true in a microwave counter. However, because of the additional front-end circuitry, the characteristics of the input signal, and the response time of the various circuits involved, the measurement speed of a microwave counter is typically longer than that of a low-frequency conventional counter. This additional acquisition time means that it is necessary to detect the microwave signal and cause the various circuits to adjust, or settle, before the measurement can take place. It takes time for a vco to scan a frequency range, and it takes a finite period of time for a phase-locked loop to settle and lock. Depending on the type of down-conversion technique used, this additional acquisition time produces a total measurement time of 100 to 400 milliseconds. This does not apply to prescaling. Prescaling requires no additional acquisition time, since it involves only frequency division.

## AM and FM Tolerance

The terms "am tolerance" and "fm tolerance" refer to the ability of a microwave counter to make frequency measurements on microwave signals which are amplitude or frequency modulated. Most microwave signals whose frequency is to be measured are accompanied by amplitude or frequency modulation of some kind. This is normal because most microwave signals are transmitting information by this modulation. In most cases, a good counter can tolerate the modulation and still produce an accurate frequency measurement. The ability of the counter to accommodate modulation and still produce an accurate measurement is reflected in the tolerance specification.

In order to provide high tolerance to amplitude modulation, many microwave counters incorporate automatic gain control (agc) circuitry. This circuitry helps to minimize the amplitude excursions of the signal. The agc circuits try to smooth out the modulation peaks and valleys and provide a signal of more constant amplitude to the counter. However, not all microwave counters contain this agc circuitry. In these counters, the am tolerance can be compensated for to some extent by appropriate adjustment of the input sensitivity. Many microwave counters have sensitivity adjustments that allow the trigger level to be adjusted, thus permitting a wide range of amplitude modulation to be accommodated. Recall that modulation is expressed in terms of a percentage from zero (no modulation) to full modulation (100%). Most microwave counters can accommodate modulation as high as 90%. The only requirement is that the trough of the am waveform be a minimum level within the normal sensitivity specification of the counter.

Frequency-modulated signals can also be measured with most microwave counters. Recall that in frequency modulation, the amplitude of the signal remains constant, but the frequency is changed in response to the modulation. Higher modulation produces greater frequency deviation from the central carrier frequency. The frequency modulation causes the instantaneous frequency of the signal to vary above and below the central carrier frequency.

Since the frequency of an fm signal is constantly changing, the frequency displayed by a counter measuring such a frequency will essentially be an average of the frequency during the gate time. If the gate time is many times longer than the period of the modulating signal, then the displayed value will be very close to the actual value. Just keep in mind that a frequency-modulated microwave signal will make it that much more difficult to provide a very accurate measurement. Most microwave counters can tolerate a certain amount of fm and give an accurate measurement. Instead of expressing the fm tolerance in terms of a percentage, it is expressed as a frequency deviation from the carrier frequency. This specification will vary from 1 MHz to 50 MHz, depending on the type of down-conversion technique used. A higher value means a better specification.

### Amplitude Discrimination

Amplitude discrimination is a specification that refers to the ability of a counter to measure the frequency of a signal while in the presence of other, lower-level signals. Most microwave counters incorporate amplitude-discrimination circuits in their design that allow them to find and capture the largest signal applied to their input. The ability of the counter to distinguish between signals whose amplitudes are similar is known as *amplitude discrimination*. This specification is typically expressed in terms of decibels (dB). This represents the difference in amplitude by which a counter can distinguish separate input signals within its frequency range. A smaller number indicates greater amplitude discrimination. Specified values vary from 2 to 30 dB, depending on the method of down conversion used.

### Signal-to-Noise Ratio

The signal-to-noise ratio (snr) refers to the ability of a counter to measure a signal in the presence of noise. This specification is usually expressed in decibels and is the ratio of the signal and noise levels that a counter will tolerate and still make a reliable measurement. The lower the decibel value, the better the counter.

## Comparison of Specifications

Depending on the type of down-conversion technique used, the specifications of microwave counters will vary. Some down-conversion techniques are better than others in some specifications. It is up to the user to determine which specifications are most important to him and, therefore, choose a counter with the down-conversion method that will provide those specifications. To help in this regard, a summary comparison chart of specifications for the various types of down-conversion techniques is given in Table 4-1.

Table 4-1. Comparison of the Three Major Methods of Down Conversion Used in Microwave Counters

| Characteristic | Heterodyne Converter | Transfer Oscillator | Harmonic Heterodyne Converter |
|---|---|---|---|
| Upper Frequency Range | 20 GHz | 23 GHz | 40 GHz |
| Measurement Speed | 150 ms | 150 ms | 350 ms |
| Accuracy | Time-Base Limited | Time-Base Limited | Time-Base Limited |
| Sensitivity/ Dynamic Range | −30 dBm/35–50 dB | −35 dBm/40 dB | −30 dBm/35–50 dB |
| Signal-to-Noise Ratio | 40 dB | 20 dB | 20 dB |
| AM Tolerance | Less Than 50% | Greater Than 90% | Greater Than 90% |
| FM Tolerance | 30–40 MHz p-p | 1–10 MHz p-p | 10–50 MHz p-p |
| Amplitude Discrimination | 4–30 dB | 2–10 dB | 2–10 dB |

## LOW-FREQUENCY COUNTERS

The measurement of low-frequency signals represents almost as much of a technical challenge as the measurement of microwave frequencies. The problem lies in achieving high-resolution measurements with frequencies below 10 kHz and the ability to measure frequencies as low as 0.001 Hz. The challenge is to provide not only the resolution and accuracy normally expected of a digital counter, but also to provide the convenience of direct digital frequency display and fast measurement speed. Low-frequency counters are more in demand today because of the wide range of applications calling for the accurate measurement of low-frequency signals. Some examples of low-frequency measurement needs are

Audio tones in telephone systems
Audio tones in radio communications equipment

Musical-instrument tuning
Speed measurement (rpm)
Power-line frequency monitoring
Geophysics
Biophysics/Biomedical measurements

This section discusses some of the techniques used to achieve the speed, convenience, accuracy, and resolution necessary in a low-frequency counter.

## Computing Counters

The measurement of low-frequency signals with any resolution is hampered by the very nature of the operation of a digital counter. Consider the measurement of the frequency of the 60-Hz power line. Using a one-second gate, any conventional counter will simply display 60. Considering the effect of the $\pm 1$-count error, the counter could read 59 or 61 Hz. This represents an error of $\pm 1$ Hz, or 1.667%.

Increased resolution can be obtained by using a 10-second gate. Now, the counter will read 60.0, or 59.9 to 60.1 Hz with the inherent $\pm 1$-count error. The measurement error is now reduced to 0.1667%. But the price we paid for this increased resolution is increased measurement time. Ten seconds is a long time for an electronic measurement, and for many applications it is simply impractical or intolerable.

This level of resolution is still far from being adequate in solving many low-frequency measurement problems. We would, of course, extend the gate time to 100 or even 1000 seconds, but that is clearly ridiculous. Just consider what the gate time would have to be to achieve 6-digit resolution at 60 Hz. At over two and one-half hours per measurement, we clearly have an impossible situation.

The solution to this problem was discussed earlier in Chapter 2. There, it was pointed out that the best way to achieve greater measurement resolution of low-frequency signals is to measure their period. In the period mode, the counter accumulates highly accurate time-base pulses for the duration of the input signal period (time for one cycle). The display reads out the period in milliseconds, microseconds, or other appropriate time units. With this approach, we clearly achieve the desired resolution and fast measurement time, but at the expense of not having a direct digital frequency readout. We must compute the reciprocal of the period (t) to get the frequency (f), or $f = 1/t$. For example, assume that we measured an input signal whose period was displayed as 16.702689 milliseconds. Computing the reciprocal, we would know that the input frequency is 59.870599 Hz.

While a common electronic calculator can quickly and easily make this calculation, it is an inconvenience to have to do so. For an occasional single measurement, this is a small price to pay. But for frequent, repetitive measurements, this added computation is clearly an inconvenience. What we need is a built-in calculator to perform this reciprocal calculation and directly display the frequency on the counter readouts. This is what a computing counter does.

Computing counters are specialty test instruments designed specifically for high-resolution low-frequency measurements. They feature a typical frequency range of 0.1 Hz to 20 MHz and 6- or 7-digit measurement resolution. At the lowest input frequency, this represents a measurement resolution of *microhertz* (1/1,000,000 Hz). Computing counters use the period measurement mode and provide an internal calculator to compute the reciprocal. Period averaging is also used on some computing counters to improve measurement accuracy further. Measurement speeds of 100 ms or less are typical.

Computing counters have been available since the late 1960s, when digital integrated circuits made it practical to develop a low-cost reciprocal computer. Computing counters of the early 1970s used PMOS LSI calculator chips which were widely available for handheld calculators. Today, modern computing counters use a microprocessor.

## Frequency-Multiplier Counters

Another method of achieving high-resolution measurement of low-frequency signals is to put a frequency multiplier ahead of a conventional counter. With this approach, the low-frequency input is multiplied to a much higher frequency which can then be measured to the desired resolution with conventional techniques. For example, a 60-Hz input would be multiplied by 1000 to 60,000 Hz, or 60 kHz. With a one-second gate, the resolution of measurement is improved several orders of magnitude.

While the concept behind this technique is simple, its implementation is more difficult. The problem is that of obtaining a frequency-multiplier circuit capable of multiplying a wide range of input signals by very high multiplication factors. The usual frequency-multiplication techniques of harmonic generation and filtering are simply inadequate for this application. It was not until the availability of the phase-locked loop (Reference XIII) that this technique became practical. Phase-locked-loop (pll) multipliers have the wide bandwidth and high multiplication capability required for this application. Multiplication factors as high as 1000 are readily obtained over a range from 0.1 Hz to 10 kHz. Such a circuit can be readily added to any counter to improve low-frequency measurement resolution. The pll multiplier can be easily switched in or out as required.

CHAPTER 5

# Counter Accessories

Most commercial counters are completely self-contained and thus require essentially no additional equipment to allow the user to make immediate use of the counter. However, there are a number of accessories available for commercial counters. These accessories are designed to enhance the counter so that it is more convenient to use. Some accessories are used to improve the performance of a counter, and others allow the counter to be used as a component in a much larger measuring system. Accessories include cables and probes, preamplifiers, prescalers, and interfaces. This chapter discusses some of the more widely used counter accessories.

## PROBES, CABLES, AND CONNECTORS

In order for a time or frequency measurement to be made with a counter, some form of interconnection must be provided between the counter and the circuit or equipment being tested. This interconnection is usually implemented by a pair of wires or a cable. In many applications, the type and length of interconnecting cable becomes a critical part of the measurement. The characteristics of the cable can greatly influence the quality of the measurement, so care should be taken in selecting and using interconnecting cables.

The type of measurement and the frequency play an important part in deciding what type of interconnecting cable should be used. For low frequencies, mainly those in the audio range (20–20,000 Hz) and below, virtually any kind of interconnecting wires can be used. One wire is used for providing a common ground, and another is the main signal-carrying wire. While such a simple pair of wires is adequate for such measurements, rarely is this combination ever used. Instead, a coaxial cable is the preferred interconnecting lead. The coaxial cable provides an insulated inner lead surrounded by the ground lead, which provides some shielding and noise immunity.

The most frequently used counter cable is a 50- or 75-ohm coaxial cable terminated in a standard BNC connector. Most counters feature the BNC-type UHF connector on the front panel. The removable cable is simply plugged into the counter connector. Fig. 5-1 shows a counter with BNC connectors and coaxial cable.

The other end of the cable can simply be the coaxial inner conductor and shield terminated with alligator clips. These clips provide a convenient means of connecting the counter to the circuit under test. Alternatively, the coaxial cable can be terminated in a probe. A probe is an insulated housing that usually contains a metallic point for connecting the counter to the circuit being measured. Often, various types of probe tips are made available. These include long, narrow, sharp pointed probes for hand application to miniature circuitry. Another frequently used probe tip is the hook or clip type, which allows the probe to be attached to a component lead. With this arrangement, it is not necessary to hold the probe tip on the component by hand. Most probes also feature a short wire with an alligator clip connected to the coaxial shield; this clip is used to ground the counter to the circuit being measured.

If the equipment being tested features output connectors, the

Courtesy Dynascan Corp.

Fig. 5-1. Digital counter with BNC input connectors.

counter cable can be terminated in an appropriate connector rather than a probe. Often, another piece of test equipment is being measured. Most test equipment uses standard BNC connectors; therefore, a cable can feature BNC connectors on each end for convenient attachment.

## Passive High-Frequency Probes

A plain piece of coaxial cable is satisfactory for use with a counter at low frequencies. If the signal to be measured is less than 1 MHz, such a simple interconnection method is satisfactory. However, at higher frequencies, the use of a plain coaxial cable becomes less desirable because of the distributed capacitance of the cable. Coaxial cables are capacitors because they consist of an inner conductor separated from the surrounding shield by an insulating medium. Most standard 50- and 75-ohm coaxial cables have a capacitance of approximately 10 to 30 pF per foot. At low frequencies, the effect of this capacitance is negligible, but at higher frequencies—1 MHz and beyond—it greatly affects any electronic measurement. When ac signals are involved, a capacitor exhibits the characteristic of reactance. The reactance ($X_c$) is inversely proportional to the frequency (f) of the ac and the capacitance (C):

$$X_c = \frac{1}{2\pi fC}$$

The higher the frequency and/or the capacitance, the lower the reactance becomes. Since the cable capacitance appears in parallel with the input to the counter, this capacitance loads, or shunts, the signal source. Fig. 5-2 shows a typical interconnection arrangement and the equivalent circuit.

The effect of the cable shunt capacitance is to attenuate the signal from the source being measured. The output impedance of the signal source and the cable capacitance plus the capacitance and internal impedance of the counter form a voltage divider. As the frequency becomes higher, the capacitive reactance becomes lower, and the signal supplied to the counter input circuits is further divided and reduced in amplitude. If the frequency is high enough, the signal can be attenuated below the point where the counter sensitivity is great enough to provide a stable reading.

The cable capacitance also forms a low-pass filter for the signal being measured. If the signal is a sine wave, then only amplitude attenuation will take place. However, if pulse or digital signals are being measured, the capacitance will effectively filter out the higher-frequency components of the signals. The result will be significant distortion of the signal. Rise and fall times will be lengthened, and most pulse and digital signals will be rounded. See Fig. 5-3. This can introduce serious measurement errors, particularly when time-interval measurements are being made.

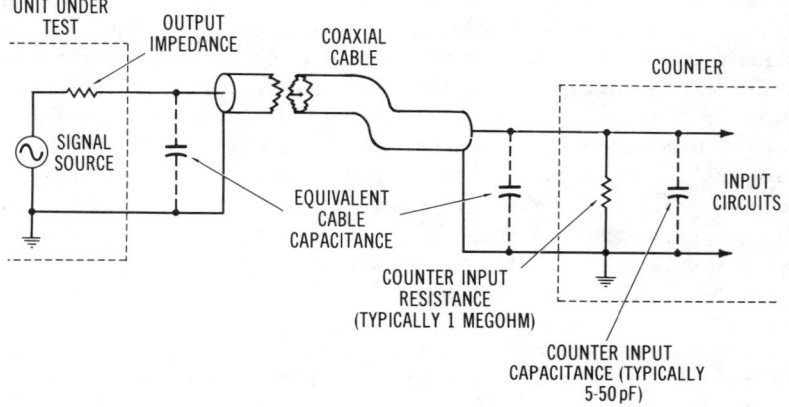

*(A) Typical arrangement.*

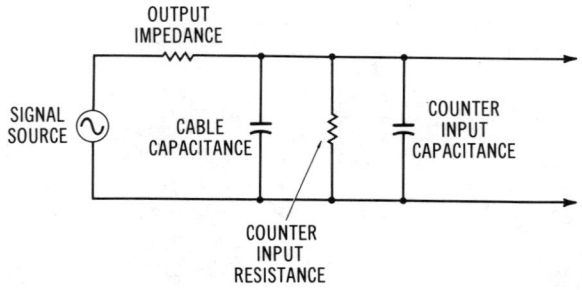

*(B) Equivalent circuit.*

Fig. 5-2. Effect of cable capacitance on counter interconnection.

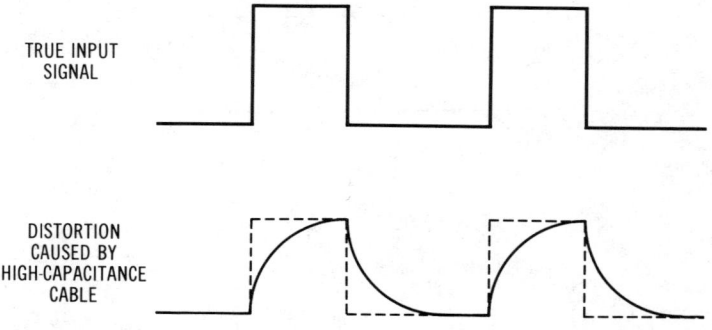

Fig. 5-3. Distortion of high-frequency pulse signal by high cable capacitance.

To get some idea of the effect of cable capacitance on signal measurement, consider the example shown in Fig. 5-4. Here a signal generator with an internal output impedance ($Z_o$) of 600 ohms is the unit under test. The frequency is 2 MHz. The signal is connected to a counter through a coaxial cable that has a capacitance of 30 pF per foot. If a 3-foot cable is used, the total capacitance contributed by the cable is 90 pF. This cable is connected to the input of a counter that has an input resistance ($R_i$) of 1 megohm. Typically, this resistance is shunted by an internal capacitance of 10 to 50 pF depending on the instrument. Assume that the shunting capacitance in this case is 10 pF. The total input impedance ($Z_i$) is 100 pF across 1 megohm. This impedance acts as the load for the signal generator.

Note that if only the resistive loading is considered (Fig. 5-4A), the voltage divider formed by the 600-ohm internal impedance and the

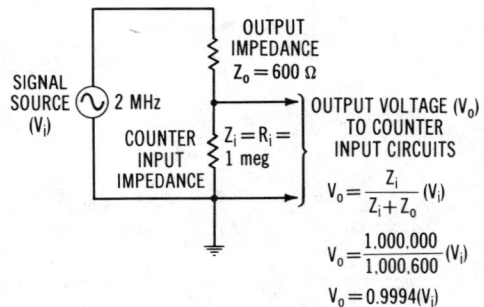

*(A) Resistance only.*

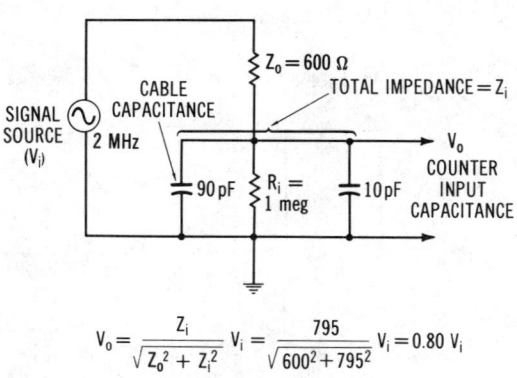

*(B) Capacitance added.*

Fig. 5-4. Effect of cable capacitance on high-frequency signals.

112

1-megohm counter impedance produces very little signal attenuation. The signal developed by the generator will appear mostly across the 1-megohm counter, and the voltage error is less than 0.1%.

However, keep in mind that a 2-MHz signal is involved. The capacitance must be considered here. The reactance of the 100-pF capacitance at 2 MHz is approximately 795 ohms. This is considerably lower than the 1-megohm counter resistance, so the total impedance ($Z_i$) of the counter input will be a reactance of approximately 795 ohms. The effect will be a considerable reduction in the amplitude of this 2-MHz signal. See Fig. 5-4B. In this example, only about 80% of the generator signal will reach the counter. Keep in mind also that this attenuation will increase with a further increase in frequency.

A partial solution to the capacitance loading problem with high-frequency signals is to use a shorter cable or a cable with lower capacitance. Special coaxial cable is available with less capacitance per foot. Typically it is larger, less flexible, and more expensive than standard cables, and, therefore, it is much less desirable. Shorter cables can also be used to reduce the capacitive loading effects. While both these methods are used, they are not totally effective, and they lead to considerable compromises and inconvenience in measurement. There are other, better techniques for reducing the capacitive loading effects.

The most common way of overcoming the capacitive loading effect of a cable is the passive attenuator probe so commonly used with oscilloscopes and other test instruments. The passive attenuator probe works on the principle that connecting two capacitors in series creates a total capacitance which is less than that of the smaller capacitor. To make a capacitive attenuator probe, a capacitor smaller in value than the total cable capacitance and input capacitance of the instrument is connected in series with the inner conductor of the cable. The result is a capacitive voltage divider, as shown in Fig. 5-5.

Assume a total cable and instrument input shunt capacitance ($C_i$) of 100 pF. We form an attenuator by connecting an 11-pF capacitor ($C_s$) in series with the input capacitance. The total capacitance ($C_T$) of this series combination is

$$C_T = \frac{C_i C_s}{C_i + C_s} = \frac{11\ (100)}{11 + 100} = \frac{1100}{111} = 9.9\ \text{pF}$$

As you can see, the total capacitance of the combination is 9.9 pF. This is considerably less than the 100-pF shunt capacitance experienced when only the cable is used. The reactance of a 9.9-pF capacitance is considerably higher at the same frequency and thus does not load the circuit under measurement as much. It is this smaller total capacitance that the circuit under test "sees" when the counter is connected to it.

While the lower overall capacitance of the attenuator probe greatly minimizes the attenuation of higher frequencies, the capacitive voltage

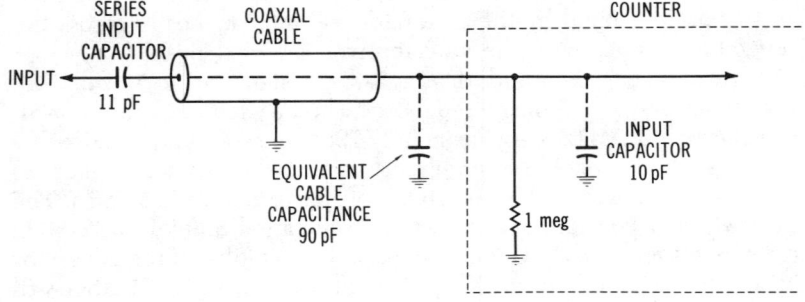

(A) *Circuit configuration.*

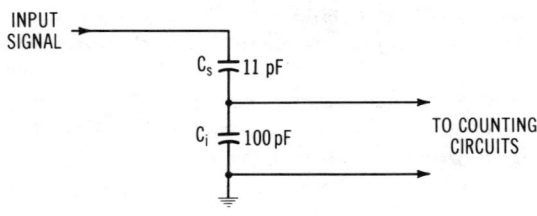

(B) *Equivalent circuit.*

Fig. 5-5. Use of series input capacitor to reduce capacitive loading effects.

divider formed does reduce the signal applied to the counter. In this case, the input voltage to the probe is reduced by a factor of 10 before being applied to the counter. The trade-off for less capacitive loading is a reduction in input voltage.

Another effect of the capacitive voltage divider is to eliminate the usefulness of the probe with dc signals. Many ac signals being measured have a dc component. Naturally, the series capacitance prevents the passage of any dc. The effect is not detrimental to ac-only measurements. However, pulse and logic signals often cannot be measured properly with a capacitive voltage-divider arrangement. This problem can be solved by connecting a resistor in parallel with the series capacitance. The value of the resistance is chosen to provide a 10-to-1 reduction in dc while the circuit capacitances provide a 10-to-1 reduction of ac. Since most counter input impedances are 1 megohm, a 9-megohm resistor connected in parallel with the 11-pF input series capacitance of the probe provides this 10-to-1 voltage division ratio. This 9-megohm resistor is in series with the 1-megohm counter input resistance. The overall result is a probe which provides a 10-megohm impedance for dc signals (Fig. 5-6). This is a much higher input

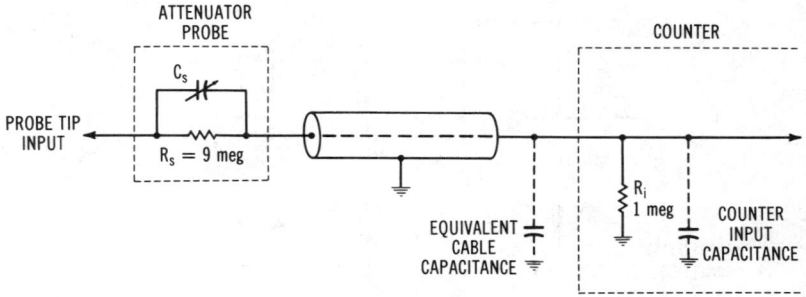

Fig. 5-6. Schematic diagram of input circuit using low-capacitance attenuator probe.

impedance than that of the counter itself and, therefore, provides less loading on the circuit under measurement. Again, the trade-off for obtaining this smaller amount of loading is the 10-to-1 attenuation factor.

The RC network created by the attenuator probe and the input capacitance and resistance can have a detrimental effect on pulse and digital signals. While the capacitive and resistive loading on the circuit or equipment under test is greatly reduced, the RC network can distort pulse signals and thus cause measurement inaccuracies. This distortion arises if the product of the probe resistance ($R_s$) and the probe capacitance ($C_s$) is *not* equal to the product of the counter input resistance ($R_i$) and the total shunt input capacitance ($C_i$). In other words, no distortion occurs if

$$R_s C_s = R_i C_i$$

If the RC products are not equal, the circuit will act like either an integrator or a differentiator, depending on which product is higher. If $R_s C_s$ is larger, the network will act like a differentiator (high-pass filter). If $R_i C_i$ is larger, the network will act like an integrator (low-pass filter). Fig. 5-7 shows the resulting effects on a square-wave input signal.

The problem in making $R_s C_s = R_i C_i$ is that the values of the cable and input capacitances are never precisely known. They vary widely with the specific counter and cable. On the other hand, $R_i$ and $R_s$ are accurately known because they are usually precision fixed resistors. The usual solution is to make $C_s$ variable. The capacitor in the attenuator probe is typically a trimmer capable of being varied over the range 5 to 20 pF. With this arrangement, $C_s$ can be adjusted to make $R_s C_s$ precisely equal to $R_i C_i$, thereby eliminating the distortion.

The procedure for adjusting the probe is to connect the probe to the counter and connect the probe tip to the output of a 1-kHz square-wave generator. Attach an oscilloscope to the counter-input end of the cable. If $R_s C_s = R_i C_i$, then the oscilloscope will show the counter input to be a

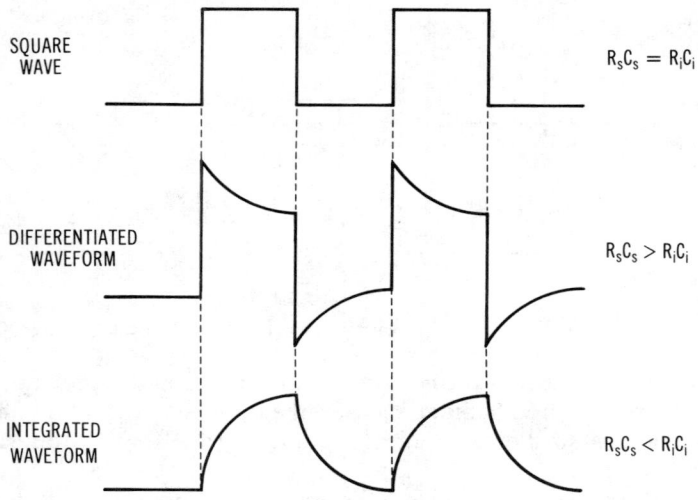

Fig. 5-7. Waveform distortion caused by improper adjustment of RC networks.

square wave. If the RC network is not properly tuned, one of the waveforms in Fig. 5-7 will result. To correct this condition, simply vary $C_s$ until a perfect square wave is noted.

Since $C_s$ is adjustable, it is used to compensate for RC network differences. For that reason, $C_s$ is often called a *compensation* or *compensating capacitor*. This capacitor is normally installed in the probe handle and is adjusted with a screwdriver. Because the probe is adjustable, it can be used with any piece of test equipment that has a 1-megohm input impedance. Most scopes and counters have such an impedance. But, since input capacitances vary widely, the probe will have to be adjusted separately for each instrument.

The use of a passive attenuator probe like that just described is highly recommended for counters. Even though such a probe does introduce a 10-to-1 attenuation, the sensitivity of most counters is good enough to offset the attenuation. If the sensitivity is not sufficient, an external preamplifier can be used. Many commercial passive attenuator probes are available. Most of them have been designed for use with oscilloscopes, but they are just as useful with digital counters. Most of these probes feature a BNC connector on one end and a convenient hand-held probe on the other. Various types of probe terminations are available for connecting the probe to the circuit under test. Fig. 5-8 shows a typical attenuator probe.

There are several variations of attenuator probes available. For example, another type of probe features 100-to-1 attenuation. Such probes feature 100-megohm input impedance with a shunt capacitance

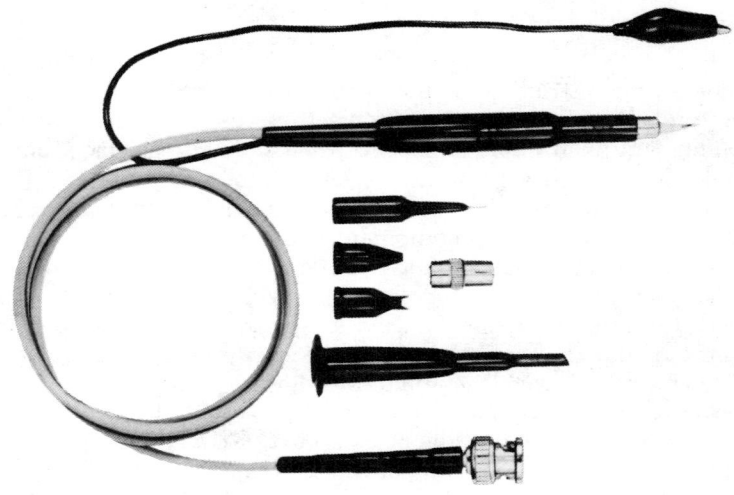

Courtesy Heath Co.
Fig. 5-8. Typical attenuator probe with accessories.

of 1 to 4 pF. This greatly reduces the loading of the circuit for high-frequency measurements. However, the 100-to-1 attenuation introduced by this probe is often a severe handicap.

Some passive probes contain a built-in switch that allows the user to select either 10-to-1 or 1-to-1 attenuation. With the switch in the 1-to-1, or × 1, position, the series capacitance and resistance are bypassed, thus eliminating the attenuator effect. With this arrangement, the input impedance of the counter and the resulting shunt capacitance of the cable are connected directly to the circuit being measured. If this is undesirable, the user can simply place the switch in the 10-to-1, or × 10, position, thus giving him a standard 10-to-1 passive attenuator probe.

Most passive probes of this type are provided with a small grounding lead. This short lead usually is terminated in an alligator clip (Fig. 5-8). The purpose of this small lead located near the probe itself is to ground the counter or other instrument to the circuit being measured. This small ground lead is connected to the shield of the coaxial cable. This ground must be connected to the circuit to provide a complete signal path. Be sure to connect it to the circuit ground; otherwise, erroneous measurement results can be obtained. Using a separate grounding wire between the counter and the circuit under test is not a satisfactory approach, because it will often introduce distortion and ringing into the circuit, particularly with pulse or high-frequency signals. For best results, it is absolutely essential that the grounding lead provided with the probe be used.

## Active Probes

Active probes are also widely used to minimize the loading effects of the measuring instrument on the circuit or instrument under test. An active probe is a probe that contains electronic circuitry and most often provides very high input impedance, low shunt capacitance, and in some cases amplification. Such electronic circuitry in the probe effectively isolates the cable capacitance and the input impedance of the instrument itself from the equipment under test.

Usually, the electronic circuitry contained in the probe is an FET follower circuit. The FET follower provides a very high input impedance, often many megohms; a 100-megohm input impedance is not uncommon. The shunt capacitance of such a probe can be as low as 1 pF. Such probes provide minimum loading regardless of the circuit impedance or the frequency involved. A typical FET-follower active probe can accommodate frequencies up to 2 GHz. The output of the FET follower is usually a low impedance and thus is not greatly affected by the shunt capacitance of the cable connecting the probe to the instrument. Recall that a voltage-follower circuit has high input impedance and very low output impedance, usually with unity gain.

Some active probes also contain amplification. They feature high input impedance and low shunt capacitance as well as gain. In effect, they give all the advantages of high impedance and low capacitive loading with no attenuation.

Active probes must have very wide bandwidth and low distortion in order not to modify or distort the signal being measured. This is particularly true in time-interval measurements on pulse signals. Whenever pulse rise time, fall time, width, and spacing are being measured, it is essential that the probes not introduce distortion or filtering effects that modify the signal and introduce measurement error. Counters designed specifically for time-interval measurements are often supplied with special matched active probes that introduce a minimum of distortion and error.

## RF Pickups

When digital counters are used for high-frequency radio measurements, occasionally a direct interconnecting cable is not desirable or necessary. In such applications, it is only necessary to connect an antenna to the counter input. The antenna is typically nothing more than a vertical whip attached to a BNC connector. Usually a telescoping antenna is used so that the length can be adjusted to approximate a quarter wavelength at the measuring frequency. Some counters use the flexible type of antenna so common in hand-held transceivers. Fig. 5-9 shows a telescoping antenna accessory for use with any counter.

Some portable handheld counters come with a built-in antenna. These counters are designed specifically for frequency measurements

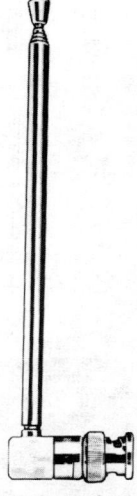

Fig. 5-9. Counter antenna accessory.

Courtesy Heath Co.

on radio equipment in the field. An example of such a counter is shown in Fig. 5-10.

To make a measurement with a counter and an antenna accessory, the counter is placed near the radiating antenna of the transmitter whose frequency is to be measured. It is best not to get too close to the transmitting antenna in order to avoid overload and counter damage. Usually, the counter should be separated from the transmitting antenna by several feet for best results.

Another often-used method of transferring rf signals to a counter is by a pickup loop. This is nothing more than a small coil of wire connected to one end of the counter cable. The coil can be made out of insulated hookup wire. It usually consists of two to four turns of wire 2 or 3 inches in diameter. Heavy wire such as No. 14 works best because it is self-supporting. Tape can be used to hold the turns of the coil together.

To use the rf pickup loop, it is simply placed near the output of the circuit to be measured. It can be looped over coils or other components so that some signal is inductively coupled into the loop. The position of the loop should be experimented with for optimum coupling and maximum input signal to the counter. Maximum signal is usually obtained when the axis of the loop is parallel to the axis of the coil to which it is coupled. Some counter manufacturers supply a coupling-loop accessory designed especially for rf measurements.

## Connectors and Related Items

As indicated earlier, most test instruments are provided with BNC connectors. This is almost universally true of digital counters. Most of

Fig. 5-10. Portable handheld counter with antenna.

Courtesy Continental Specialties Corp.

the cables and probes used with counters, therefore, use BNC connectors. There are also a variety of connector accessories that can be used with the counter and the various cables and probes. Most of these accessories provide additional convenience and flexibility in using the instrument.

One type of accessory is the *T connector* that allows more than one measuring instrument to be connected to the same circuit under test. For example, in most measuring situations it is desirable to connect an oscilloscope to the circuit under test along with the digital counter. This allows you to monitor the signal being measured. Instead of connecting a probe and cable from both the oscilloscope and the counter to the circuit, a T connector can be used. The T connector can be introduced at the counter or oscilloscope so that only one probe needs to be used. See Fig. 5-11. A variety of other such connectors allow multiple connections to be made. Not only is this practice convenient, but it also greatly

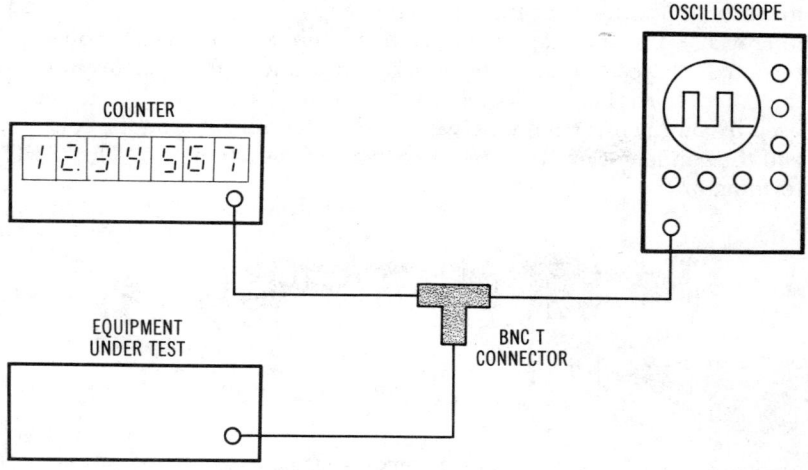

Fig. 5-11. Use of a T connector to simplify and neaten interconnections.

minimizes the mess and confusion ordinarily associated with too many interconnecting cables in a test and measurement set-up.

Another type of connector accessory is the *converter*. This device essentially accommodates two different types of connectors. For example, one type of accessory converts a standard ¾-inch banana-plug connection into BNC format (Fig. 5-12). Units for adapting type BNC to type N or type UHF are also available. Another converter accommodates the old-style UHF (PL 259) connector and converts it to BNC format. In most cases, converters are available to transform any format to another. This allows all types of cables, instruments, and equipment to be accommodated without special makeshift interconnections, which can introduce measurement errors.

Another popular connector accessory is the *terminator*, which is essentially a straight-through BNC connector that contains a nonreactive 50-ohm resistor in shunt with the circuit. In making high-frequency, particularly microwave, measurements, it is very important that cable terminators be used. In low-frequency applications, the coaxial cable acts as nothing more than a shielded interconnecting wire. However, at microwave frequencies the interconnecting cable acts as a transmission line. In order for a transmission line to perform properly, it

Fig. 5-12. A connector converter for adapting banana plugs to a BNC connector.

must be correctly terminated at the end of the cable. Since most microwave transmission lines have a 50-ohm impedance, 50-ohm terminators are sometimes necessarily introduced in order to provide the correct termination if it does not otherwise exist. Such a device is used by simply inserting it between the BNC connector on the counter and the connector on the cable. Fig. 5-13 shows a typical 50-ohm BNC terminator.

Courtesy Heath Co.

Fig. 5-13. A 50-ohm BNC cable terminator.

The usual effect of improper termination is a high vswr (voltage standing-wave ratio). See Reference XV. High vswr can cause signals of extremely high or extremely low amplitude. Low-amplitude signals may result in insufficient signal amplitude for the measurement to take place. Excessively high-amplitude signals brought about by large standing waves can cause damage to the measuring instrument.

When pulse signals are being measured, the interconnecting cable also acts as a transmission line. Improper termination results in high vswr, which, in turn, often results in signal reflections that introduce extra pulses. These extra pulses will be counted and measured along with the real pulses. The result, of course, is tremendous inaccuracy in time and frequency measurements. Terminators, therefore, are an extremely important accessory for microwave and high-frequency pulse measurements.

An accessory connector often used at high frequencies is a *connector fuse*. This accessory is nothing more than a BNC connector housing that contains a fuse element. A typical unit is shown in Fig. 5-14. The fuse will blow when a signal of excessive amplitude is applied to it. The primary advantage of this accessory is simply to protect a counter, an oscilloscope, or other measuring instrument from excessive signal amplitude. Often, the amplitude or other characteristics of the signal being measured are not known, and, therefore, damage could result if the signal is connected to the counter directly. A mismatched transmission line or one that is improperly terminated will also introduce excessively high-amplitude signals. If in doubt about the signal amplitude or vswr, use a fuse to protect the equipment.

Another accessory often associated with connectors is the *attenuator*. An attenuator is simply a resistive divider network that is used to reduce

# REFERENCE XV
## VSWR

The abbreviation vswr stands for *voltage standing-wave ratio*. The vswr is a measure of the mismatch between the *characteristic*, or *surge*, *impedance* of a transmission line and its load impedance. All transmission lines have a characteristic impedance that is a function of the physical nature of the line, such as wire or conductor size, conductor spacing, and type of dielectric insulator separating the two conductors. Most coaxial cables have a characteristic impedance in the range of 50 to 75 ohms.

Any cable or transmission line is used to transfer a signal from some ac, rf, or pulse source to a load impedance. For best results, the load impedance, $Z_o$, should be equal to the line impedance, $Z_L$, or $Z_o = Z_L$. In this way, maximum power transfer takes place, and all of the power supplied by the source is absorbed by the load.

The vswr is the ratio of $Z_L$ and $Z_o$:

$$\text{vswr} = \frac{Z_L}{Z_o} \quad \text{or} \quad \text{vswr} = \frac{Z_o}{Z_L}$$

(The larger of $Z_L$ or $Z_o$ is placed in the numerator so that vswr is a number that is unity or greater.) If $Z_o = Z_L$, then vswr = 1. Under these conditions, it is said that the line is *flat*.

If a transmission line is terminated in an impedance other than the characteristic impedance ($Z_o > Z_L$ or $Z_o < Z_L$), then maximum power transfer will not occur. All of the power delivered by the source will not be absorbed by the load. Instead, some of the signal energy will be reflected back toward the source. The reflected energy will combine (algebraically add) with the forward energy from the source. The result will be standing waves set up on the line. Standing waves represent the fixed distribution of current and voltage along a transmission line. Besides the inefficiency caused by improperly matched line and load impedances, the reflected signals create extra high or extra low voltages along the line, and these abnormal voltages can cause equipment damage or improper operation. When pulses are transmitted, the result can be pulse distortion or extra pulses. All are undesirable conditions that should be corrected by attempting to match the line and load impedances as closely as possible. The greater the mismatch, the greater the vswr and the more severe the impact will be. For example, if a 50-ohm coaxial line is terminated with a 300-ohm load, the vswr is

$$\text{vswr} = \frac{Z_o}{Z_L} = \frac{300}{50} = 6$$

This huge mismatch will usually create problems. The solution is to change the load so that it is as close to 50 ohms as possible.

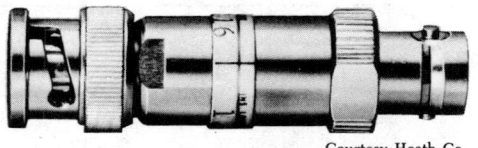

Courtesy Heath Co.

Fig. 5-14. A BNC connector fuse.

the signal amplitude. Like the terminator and fuse, the attenuator is usually housed in a BNC connector unit. Both input and output BNC connectors are provided. Usually, the attenuator is connected to the input BNC connector on the counter. The probe is then connected to the attenuator unit. Attenuation levels of 10 to 50 dB in 10-dB increments are usually available. These units are used to reduce the amplitude of a signal to prevent damage to the counter. Normally, the attenuators are provided with a constant 50-ohm input and output impedance. This allows the attenuator to serve as a terminator as well.

## PREAMPLIFIERS

A preamplifier is an external accessory used to amplify the voltage level of a signal whose frequency, period, or other characteristic is to be measured. The signal may be of extremely low level and, therefore, below the sensitivity range of the counter. The preamplifier can be used to boost the signal level into the sensitivity range of the counter.

A typical preamplifier accessory consists of a broadband linear solid-state amplifier that covers the frequency range of approximately 1 MHz to well over 1 GHz. The gain of the amplifier is typically in the 20–30-dB range. This amplifies the voltage by a constant factor of greater than 10 to 1. The preamplifier usually has input and output impedances of 50 ohms.

An external preamplifier is usually provided with BNC connectors for attaching the preamplifier to the counter input and to the various available cables and probes. The supply voltage for this external device is often supplied by a separate, internal low-voltage power supply. Some preamplifier accessories are designed to draw power directly from the counter. In this case, a dc-voltage output terminal is usually provided on the counter front panel for connection to the preamplifier.

In addition to its ability to increase the amplitude of low-level signals, the preamplifier can also be used to offset the loss introduced by a high-impedance attenuator probe. Often a 10-to-1 resistive-divider attenuator probe will be used with the counter to minimize circuit loading. While the high-impedance probe does greatly reduce the amount of resistive and capacitive loading on the circuit, the probe by its nature attenuates the signal. If the signal is a low-level one to begin with and marginally within the sensitivity range of the counter, the use of the probe can make the signal too small to be measured. In such applications, a preamplifier can be used to offset the attenuation of the probe. The preamplifier restores the original higher signal amplitude but at the same time permits the benefit of minimum loading by the probe to be realized.

## PRESCALERS

A prescaler is a frequency divider that converts a high-frequency input signal into a lower-frequency signal within the range of the counter. Prescaling is a popular technique for measurement of very-high-frequency signals. Often, the prescaler is built into the counter itself. However, external prescaling accessories are available to increase the frequency range of any low-frequency counter. A prescaler is simply an accessory that contains a preamplifier and frequency-divider circuit that divides the input frequency by a fixed ratio such as 10 or 100, thus bringing it into the measurement range of a much lower-cost low-frequency counter.

Most commercially available prescalers are designed to measure frequencies from approximately 500 MHz to 1 GHz. Of course, frequencies below these can also be accommodated. Frequency division ratios of 10 and 100 are common. With a 100-factor prescaler, a 1-GHz signal is reduced to 10 MHz, well within the range of most economy counters. A 10-to-1 division ratio reduces a 1-GHz signal to 100 MHz, which is well within the range of most medium-grade counters.

Fig. 5-15 shows a simplified block diagram of a typical prescaler accessory. It consists of a preamplifier and two high-frequency bcd decade counters used as frequency dividers. A switch is usually provided to select a division ratio of 10 or 100. The input and output impedances are typically 50 ohms for high-frequency applications. Usually, standard BNC connectors are provided on both the input and the output for connecting the prescaler to the signal under measurement and the counter itself. A typical accessory prescaler is shown in Fig. 5-16.

Most prescalers contain their own power supplies. This allows them to operate independently of the counter itself. It is not necessary to

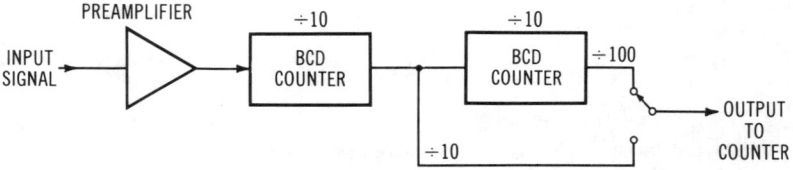

Fig. 5-15. A counter prescaler accessory.

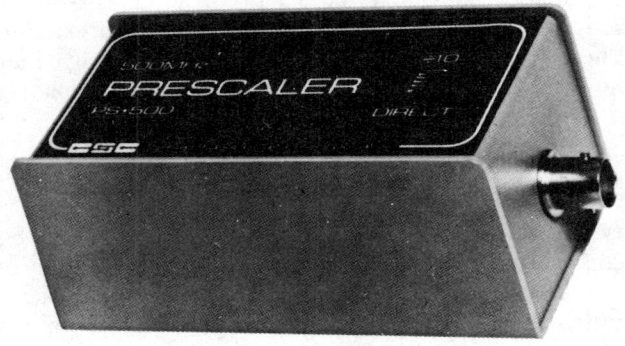

Courtesy Continental Specialties Corp.

Fig. 5-16. A 500-MHz divide-by-10 prescaler accessory.

make special modifications to the counter to provide power to the accessory scaler.

The specifications of available prescalers vary considerably. However, a typical unit may have a frequency range of 30 MHz to 1 GHz. The sensitivity is often in the range from $-40$ to $-50$ dBm. Division ratios of 10 to 1 and 100 to 1 are the most common.

When using a prescaler, remember that the input frequency is being divided by a factor of 10 or 100. Therefore, to provide the correct output reading, the display value must be multiplied by 10 or 100 as appropriate. Also remember that prescaling allows the measurement of higher frequencies but at the expense of measurement resolution. This may or may not be a disadvantage, depending on the application. The resolution limitation can be overcome by adjusting the gating period so that it is 10 or 100 times longer to offset the input prescaling factor. The longer measurement time is a reasonable trade-off for the advantage of being able to measure these higher-frequency signals.

Prescalers are one of the most popular accessories available for counters. An individual can purchase a very low-cost general-purpose counter and easily extend the frequency-measuring range later by purchasing a prescaling accessory.

## INTERFACES

Many general-purpose digital counters contain interface circuitry that allows them to be connected to external circuits and equipment. In some cases, this interface circuitry is available as an option. The purpose of the interface circuitry is to allow the counter to provide information to some external device or to allow some external device to control the counter.

A very common example of an external device is a printer that causes the value displayed on the counter readout to be printed on paper. When a counter is used for various frequency or time measurements, it is often necessary to make a series of measurements. To eliminate the need of recording such measurements manually, an external printer may be connected to the counter. Each time the counter makes a measurement, the value is displayed on the counter readouts. At the same time, the measured value is also printed. This feature is particularly valuable when it is desirable to automate the measurement process.

Counters are often used as part of a larger automatic measuring system. In such systems, it is desirable to control the counter as well as monitor its outputs. In order for the measurement process to be fully automated, the system must be able to select the counter mode, time base, and other conditions. In this way, the counter can be set up prior to making the desired measurement or changed as needed to accommodate different measurements.

An interface is the circuitry that connects the counter to any external equipment. See Fig. 5-17. It is the purpose of the interface to make the counter outputs compatible with the printer or other external device. The interface also allows the counter to be controlled by an external device. A wide variety of methods have been developed for controlling and monitoring a digital counter. In this section, we review the most popular systems.

## BCD Interface

The simplest and most widely used counter interface is one that allows the bcd outputs of the decade counting units to be monitored or recorded externally. In practice, it is not the decade-counter outputs that are directly monitored. Recall that in most counters the decade outputs are fed to storage registers before being applied to the display. It is the outputs of these storage registers that are made available to the interface. Each bcd decade output consists of four lines that carry the 8-4-2-1 bcd code for the decimal digit stored. Each of these four lines is

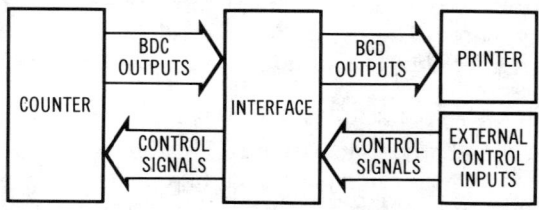

Fig. 5-17. Basic counter interface.

terminated on a rear-panel connector (Fig. 5-18). For example, if a counter contains seven decade counter sections, then there will be 28 output lines, four lines of bcd information for each decade of output information.

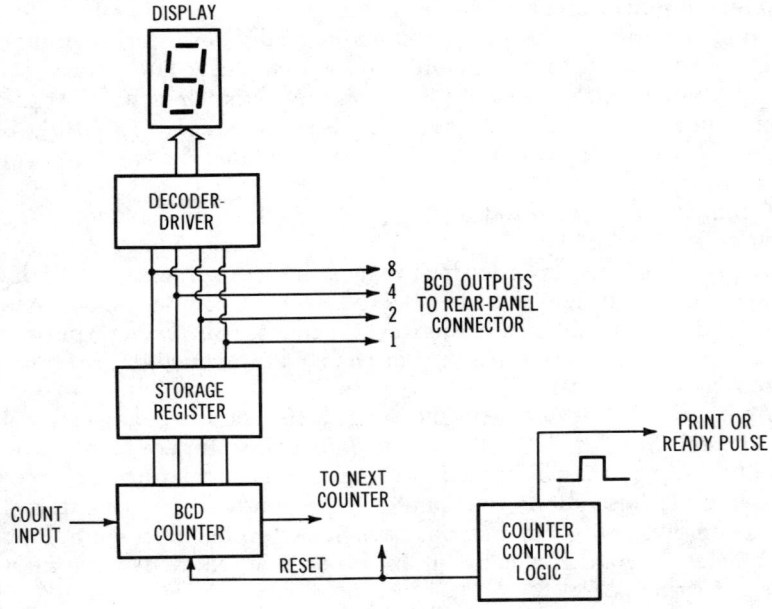

Fig. 5-18. Typical circuit configuration and counter outputs for a bcd interface.

In addition to providing a bcd output, some counters also provide a control line called "print" or "ready." This line carries a logic pulse generated by the control logic in the counter. When a measurement has been made, stored, and displayed, the counter control logic generates a pulse to the interface that signals the external device that the bcd outputs can be accepted and used.

The bcd counter outputs can be used in a variety of ways. The printer described earlier is only one common application. Sometimes an external display is used. The application may call for an external readout with larger, brighter digits. The interface contains the extra display drive circuitry.

The bcd outputs can also be fed to a computer. The computer can store the bcd data and process it as called for by the application. The role of the interface in a computer application is to make the parallel bcd data compatible with the computer. The interface will format the data in such a way that it is usable by the computer. For example, if the computer contains an eight-bit microprocessor, the interface may format the bcd data into eight-bit words or bytes. Two four-bit bcd digits can be used to make up one computer word.

Another type of interface is one that converts the parallel bcd data into serial ASCII words. (Reference XVI defines ASCII.) Here, the interface would add the extra bits to each four-bit bcd word to form the proper ASCII code. The interface may then transmit the ASCII word serially to a computer over a standard EIA RS-232-C serial i/o port. (See Reference XVII for information on RS-232 interfaces.)

The role of the interface in controlling a counter is to translate commands from some external source into signals that will change the mode, select the desired time base, or make other adjustments that might ordinarily be made from the counter front panel. For example, the control command may come from a computer in the form of a binary word or code. The interface decodes the command word and then generates the signals that set up the desired conditions in the counter.

## REFERENCE XVI

### The ASCII Code

The ASCII code is a special form of bcd code used in most computers and data-communication systems. It is a seven- or eight-bit code that is used to represent numbers, letters (both upper and lower case), special symbols (like punctuation marks), and control functions. For example, the ASCII code for the number 7 is 0110111.

The abbreviation ASCII stands for American Standard Code for Information Interchange. Most computers and microprocessors communicate with external peripheral devices via the ASCII code. In this application, the ASCII word is usually transmitted serially.

## REFERENCE XVII

### RS-232 Interface

The designation RS-232 identifies a standard for transmitting data serially in a computer or data-communication system. This standard, developed by the Electronic Industries Association (EIA), defines logic-signal voltage levels, control signals, and physical connector sizes. The EIA RS-232 logic-signal levels are nominally $+12$ volts (binary 1) and $-12$ volts (binary 0). The connector is a 25-pin unit called a D connector and contains a serial data path and various control lines. Data in the ASCII format is usually transmitted over an EIA RS-232 interface.

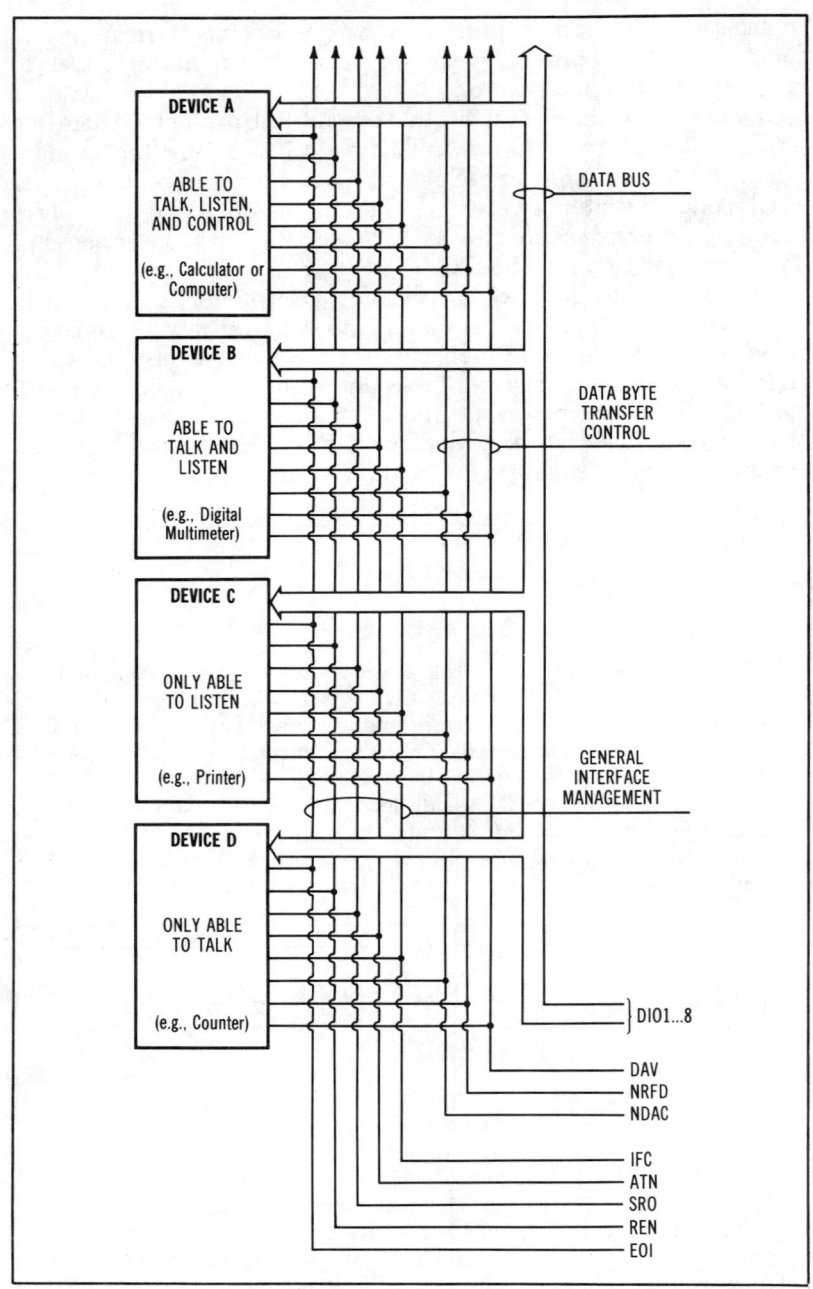

Fig. 5-19. IEEE-488 Interface and bus.

## IEEE-488 Interface

Another widely used counter interface is the IEEE-488. This is an interface standard developed by a special committee of the Institute of Electrical and Electronics Engineers. This standard defines a system for transmitting parallel data between computers and external devices. The standard defines a bus, an interface scheme, control signals and sequences, and even the mechanical details of the electrical connectors. The standard was originally developed by Hewlett-Packard, a major instrument manufacturer, for the purpose of interfacing instruments such as counters with computers. The IEEE-488 interface is most commonly used to connect devices like digital voltmeters, counters, plotters, printers, oscilloscopes, spectrum analyzers, and programmable calculators to each other or a computer. An IEEE-488 bus (also called *general purpose interface bus*—GPIB) can accommodate up to 15 such devices.

An IEEE-488 interface and bus is shown in Fig. 5-19. The bus itself contains 16 lines, eight of which are used to transmit the data as eight-bit bytes. The data transfer rate can be as high as several megabytes per second. The other eight lines are used for standard control signals that determine the direction and speed of the data transfer. The words transferred over the bus can be bcd or ASCII data or control codes.

The IEEE-488 interface usually connects to the data and address buses of a microcomputer. The control logic in the interface determines which devices will send data (talkers) or receive data (listeners) and when. The devices may communicate with one another or the computer. For example, a digital voltmeter may send data to a printer or to the computer. Or, the computer may send a control word to the counter.

The advantage of the IEEE-488 interface and bus is that it is standardized. Many counters have an IEEE-488 interface that allows them to be readily connected into a larger automated measuring system.

CHAPTER 6

# Counter Measurement Applications

You have already learned that the primary function of a digital counter is to make accurate time and frequency measurements. Throughout the previous chapters, we have given examples of how these measurements are made. In this chapter, we are going to be more specific. You will learn how to use a counter to make typical time and frequency measurements. Most of the popular applications for digital counters will be covered.

## BASIC MEASUREMENT PRINCIPLES

There are some basic rules and guidelines that should be followed to produce satisfactory measurements. If you follow these general guidelines, your specific measurement application will be successful.

**Know Your Counter**

The first important rule in making measurements with a digital counter is to know your instrument. If you are not familiar with the specifications, operating controls, and other characteristics of your counter, you run the risk of using it improperly. For example, you may try to use the counter to make measurements which it is not capable of making. You may exceed its frequency range or attempt to measure some time characteristic for which the counter is not designed. In such cases, the counter may not work at all. But more dangerous is the situation where the counter produces an incorrect output. The tendency of the individual to trust a test instrument is great. However, if used incorrectly, even the best of test instruments will give incorrect results.

In other cases, you may not be familiar enough with the counter to realize its full potential. If you have a general-purpose counter capable of making all types of measurements, you may be wasting your investment if you are not taking advantage of its many capabilities.

For these reasons, it is important to familiarize yourself with your counter. Learn everything you can about it before you begin to use it. This means reading the instruction manual. It is absolutely amazing how many people neglect reading the instruction manuals for their equipment. It is a bit tedious and time consuming, and most people want to get on with the application of the equipment rather than read the manuals. Spend as much time as you can reviewing the manual and familiarizing yourself with the capabilities and limitations of the counter. Specifically, be sure you understand its specifications. Also be sure you understand the use of all of the various controls. Do you know how to use the sensitivity and trigger-level controls? Do you understand the different modes of operation? Do you understand the implications of special modes such as period averaging? And are you familiar with the input characteristics of the counter? Are you familiar with some of the unique features of the counter, such as automatic ranging? If you can answer these questions correctly, then you are ready to use the counter.

You don't have to understand in detail how your counter operates. But you do have to know what it is capable of and what it can and cannot do. Remember, if you are not getting the expected results from your counter, go back to the instruction manual.

## Good Measurement Practices

Whenever you employ a counter to make a time or frequency measurement, you will be connecting the counter to some signal source. This will be another piece of test equipment or some other circuit or device that generates a signal. The first practice you should follow is to use proper interconnections. Probes and cables were discussed at length in a previous chapter. Be sure you follow the general guidelines given there for best results.

Second, it is good practice to observe the measurement you are making. For that reason, it is desirable to connect an oscilloscope to the counter input so that you can see the signal you are measuring. Fig. 6-1 shows the basic arrangement. If you simply connect the counter to the signal source, you are running the risk that the signal is in some way distorted or noisy and, therefore, cannot be measured accurately. By observing the signal, you can take steps to adjust the equipment itself or make compensations for signal problems by using a different probe or adjusting the counter controls. For example, you may want to change the point in the circuit at which the measurement is to be made, or you may want to adjust the signal amplitude. On the counter, you may be able to improve conditions by changing the trigger level or slope. By

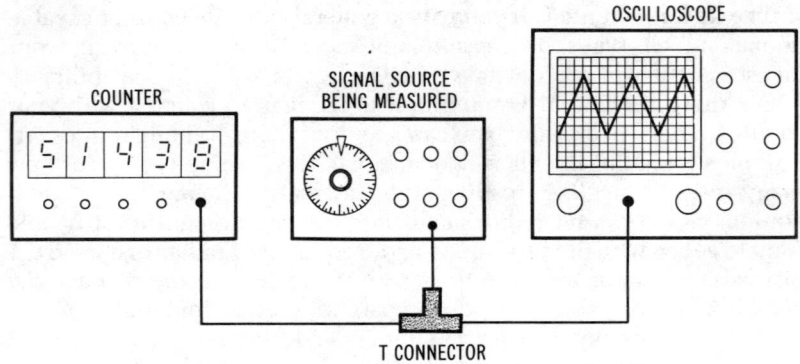

Fig. 6-1. Use of an oscilloscope to monitor the signal being measured.

looking at the signal being measured, your chances of getting the best possible measurement are greatly enhanced. For best results, always use an oscilloscope to monitor the counter input.

## FREQUENCY MEASUREMENT

Counter manufacturers have made their instruments so that even inexperienced individuals can use them effectively. Because of their broad specifications, automatic features, and customer-proof characteristics, counters are one of the easiest pieces of test equipment to use. Frequency measurement with a counter is as simple as connecting the counter to the signal source. This is particularly true of counters designed to measure frequency only (Fig. 6-2). The result, in most cases, is a virtually instantaneous and accurate frequency measurement.

But, while modern digital counters are good, there are some simple rules and procedures that can be used to ensure good measurements. These are particularly applicable when the signals being measured push the specifications of the counter to the limit. Signals whose frequencies approach the upper limit of the counter can cause trouble. Very low-level signals and signals which contain a lot of noise can also make the measurement difficult. With a little care and attention, however, even these signals can be measured.

### Connecting the Signal Source

The first step in making a frequency measurement is to connect the counter to the instrument, circuit, or equipment generating the signal whose frequency is to be measured. This is usually done with a coaxial cable terminated in a BNC connector at the counter. The other end of the cable usually has a probe, alligator clips, or a BNC connector.

Courtesy Philips Test & Measuring Instruments Inc.

Fig. 6-2. Counters designed for frequency measurement only.

In making measurements of low to medium-high frequencies, alligator clips and clip-on probes provide great convenience when working with circuits. A clip-on probe can be quickly and easily moved from one point in a circuit to another. Alligator clips are also convenient but are larger and less reliable. Probes and alligator clips are typically satisfactory for signals up to about 100 MHz. Beyond that, BNC or other standard connectors should be used for best results.

When a cable with a probe or alligator clips is used, proper grounding is essential for best measurements. Normally, two alligator clips are used on the coaxial cable, one connected to the center conductor of the cable and the other connected to the shielded braid. It is essential that both clips be connected to the apropriate points in the circuit. The shielded braid, or ground connection, should be attached to the commmon ground of the circuit. The center conductor of the cable is then connected to the signal source.

When an oscilloscope probe is used for making a connection, the clip-on probe tip is usually attached to the signal source point. However, a ground connection is also necessary. Most oscilloscope probes are provided with a short wire lead and clip which are used for making this ground connection. The short ground wire is usually a part of the probe assembly. The ground clip should be attached to the common ground near the signal source.

Proper grounding is extremely important, particularly in high-frequency and digital measurements. If the ground connection is not properly made, noise and ringing will be introduced (Fig. 6-3). These can introduce false triggering and measurement errors. Often there is

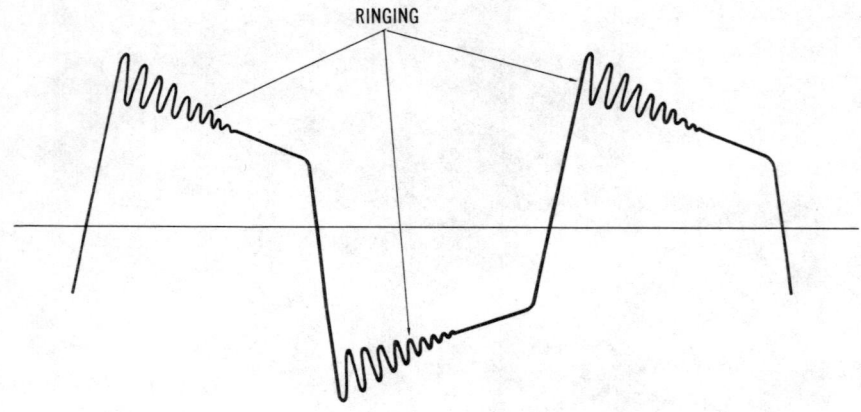

Fig. 6-3. Ringing in a pulse signal caused by improper grounding.

the temptation to use a separate ground between the counter and the equipment under test. This type of ground connection should not be used in lieu of the ground connection provided through the cable and probe because it can introduce noise and ringing.

When frequencies in the vhf range and beyond are being measured, it is best to use a cable that is properly terminated in a connector. The equipment or instrument whose signal is being measured typically has a matching connector. Connectors provide the best interconnection at these frequencies; problems with standing waves and reflections can be exaggerated with improper interconnections and terminations. In very-high-frequency measurements, terminators may be needed to eliminate high vswr and the resulting problems.

At radio frequencies, direct connection may not be necessary. By using inductive coupling or an antenna, sufficient signal can often be picked up from the signal source to allow a proper measurement. For example, when rf measurements are made in a transmitter, a "sniffer" coil can be used to couple the rf energy to the counter. The sniffer coil is simply three or four turns of heavy (No. 12 or No. 14) insulated wire wound into a coil of 2-inch diameter. This coil is connected to the coaxial cable from the counter. The sniffer coil can be moved around in the equipment adjacent to the various inductors so that mutual inductance is established. The rf signal will be coupled into the sniffer coil, translated into an induced voltage, and sent to the counter for measurement.

It is absolutely essential that a sniffer coil be fully insulated to avoid coming into direct contact with circuits being measured. Also, if high voltages are present in the equipment, extreme care should be taken to avoid shock. In any case, the sniffer coil is moved around and its position experimented with until a sufficient amount of signal is picked up and coupled to the counter. Typically, the best coupling results when the axis of the sniffer coil is parallel to the axis of another coil in the circuit. The coupling is usually sufficient when the reading on the display is stable. If the display reading fluctuates or changes erratically, the signal amplitude is probably insufficient to cause reliable triggering. The position of the coil should be readjusted until a stable reading results.

An antenna can be connected to the counter to perform frequency measurements. Many manufacturers offer a special antenna accessory designed specifically for the counter. It is typically terminated in a BNC connector and designed to mount vertically on the front or rear of the counter. This arrangement is widely used for making vhf and uhf frequency measurements. It is well suited for troubleshooting, servicing, and adjusting mobile radio transmitters and handheld transceivers whose signal typically radiates from a vertical antenna. In order to ensure best results, the counter must usually be placed close to the radiating antenna. A battery-operated counter may be needed for outside measurements on a mobile radio. Usually, a distance of several feet is adequate to provide sufficient input to the counter for reliable frequency measurement. It is normal to have to experiment with the distance between the counter antenna and the radiating antenna in order to obtain a stable readout.

**Counter Adjustments**

Many frequency counters do not require adjustment. Other, more flexible and sophisticated instruments do have several controls that require setting.

The first control to be set selects the frequency-measuring function. Such a control may not exist on some counters designed for frequency measurement only (Fig. 6-4). However, on versatile laboratory counters, some form of switch or push button is ordinarily used to select the frequency-measuring mode.

Next, the proper time base should be selected. The time-base frequency determines the resolution of the measurement. The longest gate time giving a full display should be selected for best results. Many counters have an auto-ranging feature that automatically optimizes the selection of the time base to ensure maximum resolution.

The only other controls that may need some adjustment are the trigger level and the sensitivity, if they exist on your counter. The trigger-level control selects the point on the input waveform at which a trigger pulse is generated to increment the counter. The trigger-level control may be set to zero so that on ac signals the triggering occurs at the zero-crossing point. Most trigger-level controls can also be set to either positive or negative triggering thresholds. This allows the counter to be triggered on positive-only or negative-only signals.

Courtesy Hewlett-Packard

Fig. 6-4. Examples of frequency-only counters with a minimum of controls.

The trigger-level control gives the counter great flexibility, particularly in measuring highly distorted signals or signals containing noise. If the trigger-level control is improperly set, false readings can be obtained. This is particularly true of signals that contain ringing or noise. Unusually shaped or highly distorted signals can also give false indications. However, by adjusting the trigger-level control, often the problem can be overcome and an accurate measurement of frequency can be made. Some typical examples are shown in Fig. 6-5; in these situations, incorrect triggering can be avoided by setting the trigger level to zero.

You can often determine that the signal being measured contains noise or distortion by the fact that the frequency readout is not stable but jumps around over a wide range. Also, you will usually have an idea of what the frequency readout should be. Any deviations will indicate a possible problem. In a case where high-frequency noise or ringing occurs, higher frequencies may be displayed, giving an indication of the nature of the trouble.

While an unstable display is an indicator of trouble, this trouble often can be spotted ahead of time and avoided by always using an oscilloscope to monitor the signal. In this way, you can quickly locate and anticipate

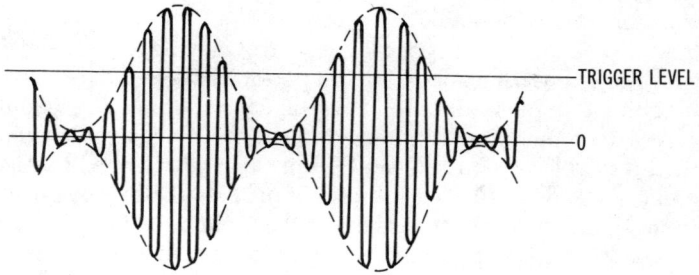

*(A) Amplitude-modulated signal.*

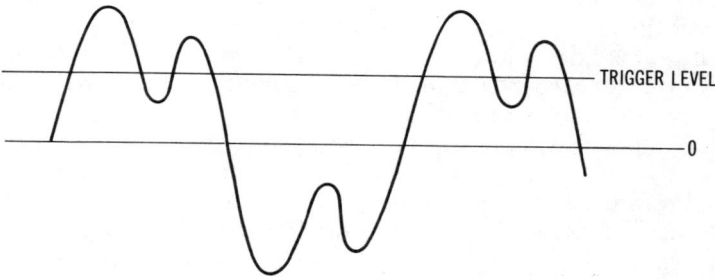

*(B) Signal with severe distortion.*

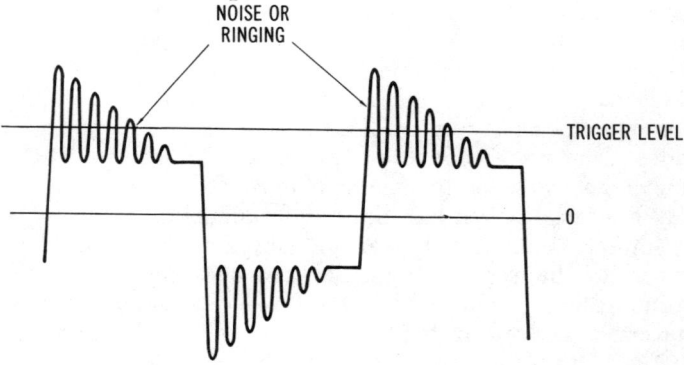

*(C) Signal with noise or ringing.*

Fig. 6-5. Examples of signals that can cause improper triggering and false frequency measurement.

the problem and make the necessary adjustments in trigger level to provide for a correct reading.

On some counters the trigger-level control may be called a sensitivity control. Regardless of its name, it still performs the same function, that of selecting the point on the input signal at which the counter is triggered. In some older counters, the sensitivity control is simply an attenuator or volume control. Most often, such a control will be labeled "attenuator." Attenuators are sometimes provided to reduce the level of the input signal to prevent overdriving the input circuits of the counter; such overdriving can produce erratic readings. In earlier-generation counters, the sensitivity or attenuator control actually adjusted the gain of the input amplifier circuit. This allowed the input circuits to accommodate a wide range of input signals. However, today most counters contain versatile input circuits with great dynamic range and overvoltage protection that virtually assure proper measurement of any signal amplitude within the sensitivity range of the counter. Nevertheless, if your counter has a sensitivity control, it should be adjusted for the most reliable triggering.

Once the controls are set and the counter is properly connected to the signal source, the display will show the frequency. At this time, you can either record the reading as required or make adjustments on the equipment as necessary to change the frequency to a desired figure. Some counters provide a hold control that allows you to store the frequency measurement so that the equipment under test can be turned off. For example, to measure the output frequency of a handheld transceiver, the push-to-talk switch is depressed, and the frequency appears on the counter display. A hold or store button on the counter is then depressed to store the reading. The handheld transceiver can then be turned off and laid aside.

**Frequency-Ratio Measurements**

One special form of frequency measurement with a counter is the frequency-ratio measurement. This is a special frequency mode that provides a means of comparing two input frequencies. That is, the frequencies of two separate signal sources are fed to the counter, and the display indicates the ratio of the two frequencies.

As you saw in another chapter, the ratio mode is almost identical in operation to the regular frequency-measurement mode. The only difference is that the internal counter time base is disconnected and the second signal is used in its place. One input signal is applied to the normal counter input, while the other is connected to the external-time-base input. As a result, the ratio of the two signal frequencies is displayed on the counter readout. For example, if 7.5 MHz is applied to the regular counter input and 15 kHz is applied to the external-time-base input, the display will read 7,500,000/15,000 = 500.

If a frequency-ratio measurement is to be made, you should first determine that the counter is capable of this measurement. Next, the ratio mode should be selected. This is usually a separate push button or switch position. On some counters, the frequency mode is the correct switch setting; in this case, a separate time-base switch is used to disconnect the internal time base so that an external input signal can be applied instead. This input is usually made through a front-panel, or sometimes a rear-panel, BNC connector.

In order to provide a high resolution of ratio measurement, the input signal which replaces the time base is often divided in frequency by the internal time-base divider. This allows a longer gating period for one input signal and thus improve the resolution. The factor by which the second input signal is divided, usually some power of ten (10, 100, 1000, etc.), can often be selected by using time-base-selection controls on the front of the counter. The time-base divider allows the greatest number of display digits to be filled for best resolution. Keep in mind, however, when selecting the division ratio that some counters may not provide for the correct positioning of the decimal point. In such cases, it will be necessary to divide the display reading by the appropriate time-base figure (move the decimal point the proper number of places to the left) in order to get the true ratio reading.

## TIME MEASUREMENTS

Many counters are designed so that their main measurement function is frequency. Such counters cannot make time measurements as such. Other, more flexible counters, particularly those designed for laboratory work, can perform both frequency and time measurements. Fig. 6-6 shows such a universal timer/counter. There are even special counters designed specifically for time-interval measurements. An example is shown in Fig. 6-7.

There are many different types of time measurements that counters can make. Most counters capable of time measurements can determine the period of an input signal. Period measurement is used instead of frequency measurement when it is desired to know the time that it takes for one cycle of the input signal to occur. Period measurements are also used to provide high-resolution low-frequency measurements. To get improved resolution at low frequencies (those below about 200 Hz), long gate periods are required. Even with a long gate time, sufficient resolution often cannot be obtained for the desired application. In such cases, measurement resolution can be greatly improved by using the period mode. The period mode and time base are selected, and the measurement is made. Of course, it is then necessary to compute the actual frequency. Some special counters incorporate computing circuits that make the time-to-frequency computation and display the result

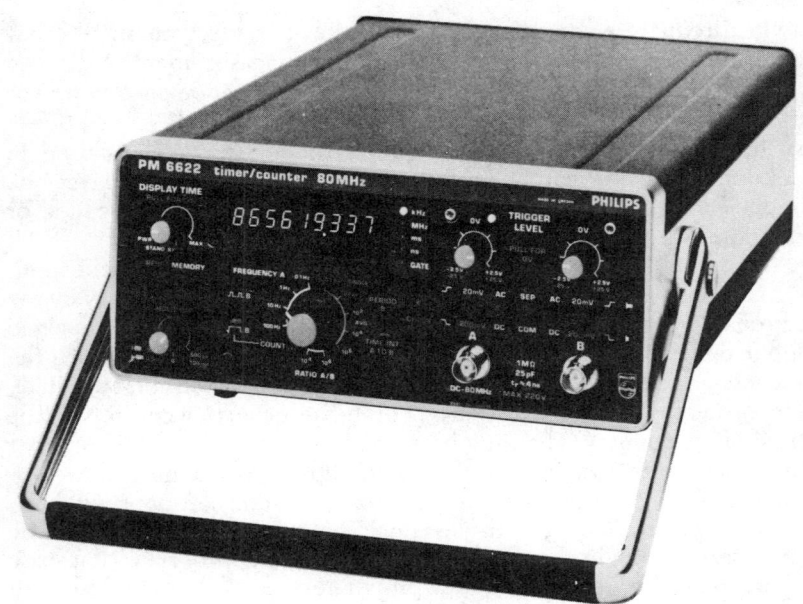

Courtesy Philips Test & Measuring Instruments Inc.

**Fig. 6-6. A universal timer/counter.**

directly in hertz. Otherwise, the frequency can be quickly calculated with an electronic calculator (f = 1/t).

Some counters also contain a period-averaging mode. Improved accuracy and resolution of period measurement can be obtained by making multiple period measurements. In this counter mode, a number of periods, usually a multiple of 10, are counted instead of a single period. The length of time required to make the measurement is, of

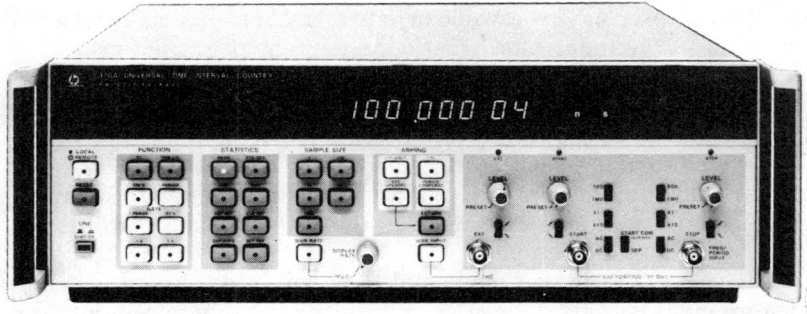

Courtesy Hewlett-Packard

**Fig. 6-7. A general-purpose time-interval counter.**

course, increased, but the resulting improvement in measuring accuracy is usually worth the longer time required.

Beyond the period and period-averaging modes, most counters are not capable of making additional time measurements. Yet there are some counters that contain special internal circuits that allow virtually any type of time-interval measurement to be made. As you recall, these counters usually contain two separate input channels, each with its own trigger-level and slope controls. This allows two independent signals to open and close the main gate in the counter. The two channels can also be connected together so that they can share a common input. This allows a variety of measurements to be made on the input signal, since there are two channels whose slope and trigger levels can be independently adjusted. Some of the pulse measurements that can be made with the time-interval measurement mode of a counter are pulse width, pulse spacing, duty cycle, rise/fall time, propagation delay, and phase shift.

In the following paragraphs guidelines for setting up the counter to perform the various time-interval measurements are given. Keep in mind that each counter is different with respect to how it makes such measurements. Therefore, it is essential that you refer to your instruction manual to be sure that each measurement is set up properly.

## Pulse Measurements

Some of the pulse characteristics that must often be measured are explained in Reference XVIII. Measurements of these pulse characteristics require a special counter that has two separate and independent input channels. Each input channel should have the necessary gain, attenuation, slope, and trigger-level controls that are characteristic of most single-channel counters. These two input trigger channels are used to open and close the main gate of the counter. One input channel opens the gate to allow clock pulses from the time base to be accumulated in the main counter. The other input channel closes the main gate and stops the count. With two separate input channels, time intervals between two separate input signals can be measured. In addition, most counters that contain two independent input channels can be reconfigured so that the two channels can be used to monitor a single input signal. In other words, a single input can be applied to both input channels at the same time. A variety of different time measurements can be made with this arrangement. By simply selecting the triggering amplitude and the polarity of the triggering slope, all of the pulse characteristics explained in Reference XVIII can be measured.

To be sure you understand this, the slope and trigger-level characteristics of these input channels should be described. First, slope refers to the direction of change of amplitude. A positive slope, for example, indicates a rising voltage or current. A negative slope

# REFERENCE XVIII

## Pulse Characteristics

Pulse signals are the repetitive rectangular voltages generated by switching circuits or digital logic elements. Unlike continuous analog signals, pulses begin and end abruptly as voltages are switched on and off. As a result, the pulses have many special characteristics. In dealing with pulse signals, it is usually necessary to measure these various characteristics. The characteristics of pulses are illustrated in Fig. XVIII-1 and explained below.

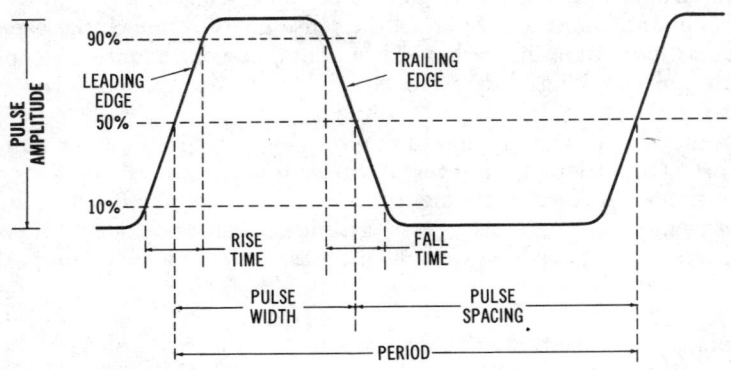

Fig. XVIII-1. Common pulse characteristics.

*Pulse Width*—The time interval between the 50% amplitude points on the leading and trailing edges of the pulse.

*Pulse Spacing*—Also known as *pulse separation*. The time interval between the 50% amplitude points on the trailing edge of one pulse and the leading edge of the next pulse.

*Period*—The time for one pulse cycle; pulse width plus pulse spacing. Reciprocal of the pulse repetition rate (prr).

*Duty Cycle*—The ratio of the pulse width to the period expressed as a percentage:

$$\text{Duty Cycle} = \frac{\text{Pulse Width}}{\text{Period}} \times 100\%$$

*Rise Time*—The time interval between the 10% and 90% amplitude points on the leading edge of the pulse.

*Fall time*—The time interval between the 90% and 10% amplitude points on the trailing edge of the pulse.

represents a decreasing voltage or current. Most input-signal conditioning channels allow the selection of either a positive or negative slope. When making this selection, you are reconfiguring the input-channel circuit so that it triggers on either a positive-going or negative-going input. If a positive slope is chosen, each time a rising input voltage occurs, a trigger pulse will be generated. Typically, this trigger pulse will then either open or close the main gate. If a negative slope is selected, a trigger pulse will be generated each time a negative-going signal is detected.

In addition to the slope, the level of the triggering can also be adjusted on most timing counters. Most input circuits consist of a comparator that determines the level which the input signal must achieve to generate a trigger pulse. If this level is made adjustable, then great flexibility is obtained in making various pulse measurements. Most level adjustments can be either positive, negative, or zero. If the level control is set for triggering at a zero level, a trigger pulse will be generated each time an input signal passes through zero. Keep in mind that the slope selection is also a factor.

Fig. 6-8 shows the internal counter circuitry involved in time-interval measurements. The input circuits have independent slope and trigger-level adjustments. The common switch allows one input signal to be applied to both input channels. Now let us see how a counter with two separate input channels can be used to make various time-interval measurements.

*Pulse Width*—To measure pulse width, the two independent inputs are connected together so that the signal is simultaneously applied to both input channels. Next, a positive slope is selected for the input channel that opens the main gate, and a negative slope is selected for the channel that closes the main gate. Finally, the trigger-level inputs are set to the 50% amplitude point of the input pulse. This is done in a

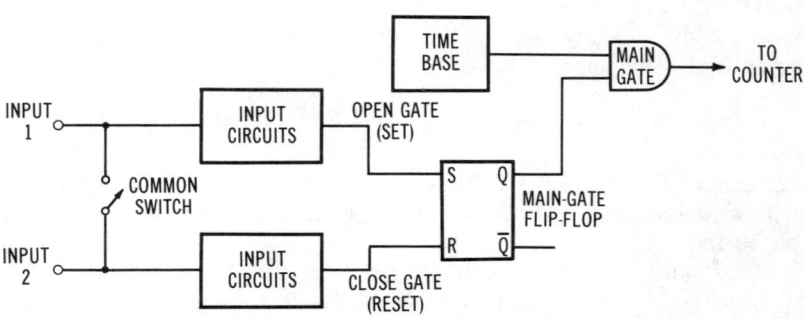

Fig. 6-8. Counter circuits involved in time-interval measurements.

145

variety of ways. On some less sophisticated counters, the 50% amplitude point has to be determined by experimentation or additional independent measurements. For example, an oscilloscope can be used to measure the amplitude of the input pulse and adjust the trigger level to the 50% amplitude point. Other counters have input trigger-level-control calibrations which make the adjustment of the trigger level easy. On some sophisticated counters, special input probes and trigger circuitry are available to allow the trigger level to be very accurately set with input thumbwheel switches or a keyboard.

Once the slope and trigger levels have been adjusted, the counter must also be set to the time-interval mode. This allows the counter to count pulses from the time base. The time-base frequency should be selected to achieve maximum measurement resolution. With this arrangment, the input-signal pulse width will be measured. When the input pulse goes positive and reaches the 50% amplitude point, the first input channel creates a trigger that opens the main gate (sets the main-gate flip-flop in Fig. 6-8, thus enabling the main gate). Pulses are then accumulated in the main counting unit. When the 50% amplitude point on the trailing edge of the input signal occurs, another trigger is generated to close the main gate (resets main-gate flip-flop in Fig. 6-8 thereby inhibiting the main gate). The resulting time interval will be shown on the counter display.

*Pulse Spacing*—Pulse spacing can be measured in the same way that pulse width was measured. The 50% amplitude point must be determined for proper measurement. The only difference between this measurement and the pulse-width mode is the selection of the trigger slope. For this measurement, the input channel that creates the trigger to open the main gate must be set to a negative slope. The channel that creates the trigger to close the main gate is set to a positive slope. With this arrangement, the counter will measure the time interval between the two pusles. See Fig. XVIII-1.

*Duty Cycle*—Duty cycle is the ratio, expressed as a percentage, of the pulse width to the period of the pulse train. Since it is possible to measure the pulse width and the period, the duty cycle can be easily computed. The formula for duty cycle is

$$\text{Duty Cycle} = \frac{\text{Pulse Width}}{\text{Period}} \times 100\%$$

For example, if the pulse width is 50 nanoseconds and the period is 150 nanoseconds, the duty cycle of this particular pulse train is calculated to be 50/150 × 100 = 33%.

Most counters cannot compute or display the duty cycle of a pulse train directly. The counter must be used to make independent pulse-width and period measurements. Once this information is obtained, the duty cycle can be computed on an electronic calculator.

Some special computing counters allow automatic measurement, computation, and display of duty cycle.

*Rise and Fall Times*—Rise and fall times can also be measured with a counter that has two independent input channels. To measure rise time, the input is applied to both channels simultaneously, and the two input channels are both set to positive slopes. The input channel that opens the main gate should be set to the 10% amplitude point of the input signal, while the channel that creates the trigger to close the main gate is set for the 90% amplitude level. With this arrangement, the main gate will be opened at the 10% amplitude level and closed at the 90% amplitude level. Timing pulses will be counted, and the rise time of the pulse will be indicated.

For a fall-time measurement, a negative slope is selected for both input channels. The channel that creates the trigger for opening the main gate is set at the 90% amplitude level, and the trigger channel that creates the trigger for closing the main gate is set at the 10% amplitude level. With this arrangement, the pulse fall time will be displayed on the counter.

Most rise and fall times are extremely short periods of time. For modern digital circuitry, pulse rise and fall times are in the low nanosecond region. Therefore, to measure such rise and fall times accurately, a counter with an extremely high-frequency time base is required. Most counters simply do not have a high enough time-base frequency to provide accurate measurement of rise and fall times of high-speed logic circuits. On low-frequency signals with longer rise and fall times, more accurate measurements are possible. For this reason, rise and fall times must sometimes be measured on an oscilloscope. While the measurement will not be exact, the oscilloscope will give a reasonably accurate measurement when the counter may not.

**Propagation Delay**

Propagation delay is the time delay, or time shift, that occurs in a digital logic circuit between the input and output level transitions. Refer to Fig. 6-9. When a logic signal occurs at the input of a digital circuit, it typically causes an output signal to be generated. However, because of the time delay generated by the internal circuitry, there is a finite interval between the time that the input signal occurs and the time that the output signal occurs. This propagation delay is an important characteristic in digital circuitry, because it limits the upper frequency of operation. In addition, propagation delays can cause improper operation and false triggering.

It is often necessary to measure the propagation delay in a digital circuit. A dual-channel oscilloscope can be used for this purpose. However, some very-high-frequency counters may be used. In this application, the two independent channels are separated. The input

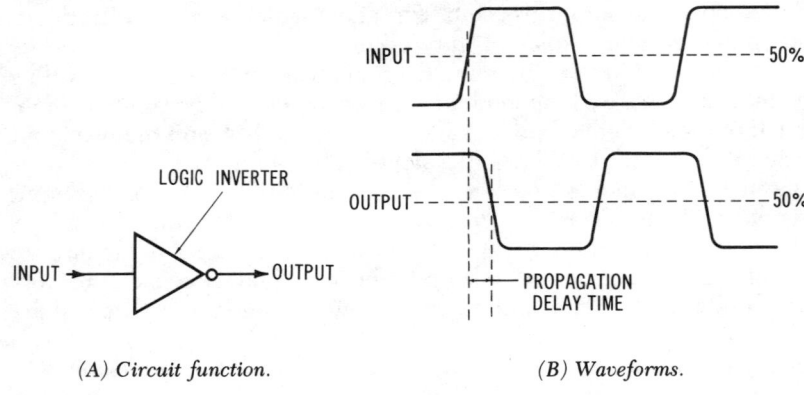

Fig. 6-9. Propagation delay in a digital circuit.

channel that opens the main gate is connected to the input of the logic circuit. The output of the logic circuit is connected to the other input channel. The slope selection for the two input channels will depend on the polarities of the logic signals and whether or not logic inversion is involved between the input and output. For example, measuring the propagation delay of a logic inverter as in Fig. 6-9 requires a positive-going input slope and a negative-going output slope.

Again, we should point out that propagation delay is a very short period of time. Propagation delays of several nanoseconds are typical in high-speed circuits. In lower-speed circuits, propagation delays as high as 1 microsecond are possible. In order to get an accurate, high-resolution measurement of propagation delay, a counter with an extremely high-frequency time base is required. One that generates a pulse-interval of 1 nanosecond or less will give the best result.

Since most counters cannot be used for measuring propagation delay because of the upper frequency limit of the time base, other methods of measurement are necessary. This is particularly true when it is desired to determine the propagation delay of a single logic element. In this case, an average propagation-delay value can be obtained for a typical circuit. This is done by connecting a series of logic elements into a loop forming a ring oscillator (Fig. 6-10). Typically five to nine (some odd number) inverters, logic gates, or other elements are cascaded and the loop closed. This circuit will then oscillate at a frequency determined by the propagation delay of the circuits. To determine the propagation delay, all that is necessary is to measure the frequency of this ring oscillator accurately. The frequency mode of your counter is used for this purpose. Assume that the frequency of oscillation is found to be 25 MHz. The period of this frequency is then determined. A frequency of 25 MHz represents a period of 40 nanoseconds. Since there are five

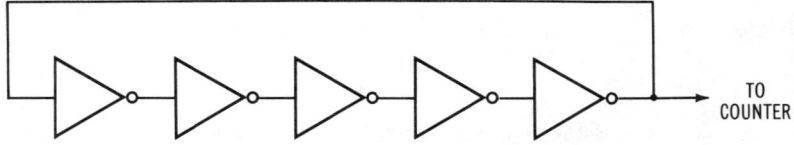

Fig. 6-10. A ring oscillator made with logic inverters to measure propagation delay.

circuits connected in series, the average propagation delay of each is 40/5 = 8 nanoseconds.

## Phase Shift

Phase shift is the time difference between two separate signals of the same frequency. See Reference XIX. While phase shift is normally expressed in degrees or radians, in reality phase shift is a time interval. Using the time-interval measurement mode, most counters can measure phase shift. The two independent input signals whose phase difference is to be determined must have the same frequency. For example, one may be the input to a filter and the other the output. Each of the signals is applied to one of the two counter input channels. Next, the slope and trigger-level adjustments are made. These will depend on the desired measurement points on the sine-wave signal. Typically, both input channels will be set for either positive or negative slope. The trigger-level setting is particularly important in a phase measurement if the amplitudes of the two signals involved are different. If the trigger level is set to the same value for signals of different amplitudes, a tremendous error in phase-shift measurement will occur. To overcome this problem, the two input-signal amplitudes should be adjusted until they are identical, or the trigger level on each signal should be the same percentage of the amplitude. If continuously variable input gain controls or attenuators are available in the counter, the signal amplitudes can be set to the desired level.

An easier way to solve this problem is simply to set the trigger point to zero for both input channels. Since both of the sine waves will pass through the zero point, the phase measurement becomes totally independent of input-signal amplitude.

With the slope and amplitude controls set, the phase measurement can begin. The counter will actually measure the time interval between identical points on the input waveforms. This measurement provides only the time difference. To compute that phase, the time difference must be compared to the period of the input signal. This means that a separate period measurement must be made. Once the time-shift and period measurements are available, the phase shift in degrees can be computed with the formula

# REFERENCE XIX

## Phase Shift

Phase shift is the time separation between two signals of the same frequency. See Fig. XIX-1. Each cycle of a sinusoidal signal is said to have 360 degrees or $2\pi$ radians. Phase shift between two signals of the same frequency is expressed in degrees or radians rather than units of time. The phase shift is computed by determining the ratio of the time shift to the period and multiplying by the number of degrees or radians in one cycle.

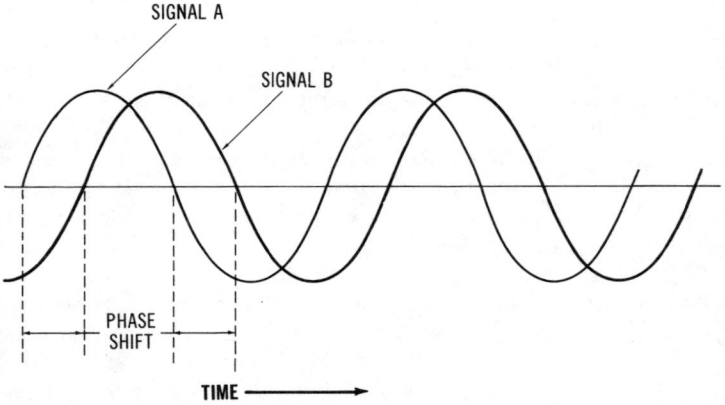

Fig. XIX-1. Phase shift between signals.

$$\text{Phase Shift} = \frac{\text{Time Shift}}{\text{Period}} \times 360°$$

$$\text{Phase Shift} = \frac{\text{Time Shift}}{\text{Period}} \times 2\pi \text{ radians}$$

Phase shift is caused by reactive (capacitive or inductive) elements in a circuit. Almost any linear circuit, passive or active, that contains inductance or capacitance will introduce phase shift. This means that the circuit output will be shifted in phase from the input. The output signal is said to either lead (be ahead of in time) or lag (be behind in time) the input. In Fig. XIX-1, signal A leads signal B because it occurs earlier in time. On the other hand, signal B lags signal A because it occurs later.

There are two special cases of phase shift. If there is no phase shift, then the maximum, minimum, and zero-crossing points of the two signals occur at the same times. The signals are said to be *in phase*; the phase shift

*Continued on next page.*

REFERENCE XIX (cont.)

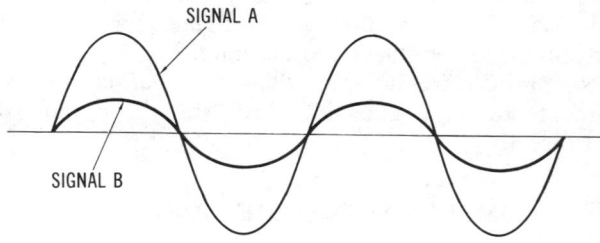

Fig. XIX-2. Signals in phase.

is zero (Fig. XIX-2). If the zero-crossing points of the two signals occur at the same time, but the maximum point of one occurs at the minimum point of the other and vice versa, the two signals are said to be 180° out of phase (Fig. XIX-3). Sometimes this condition is referred to as *phase inversion*.

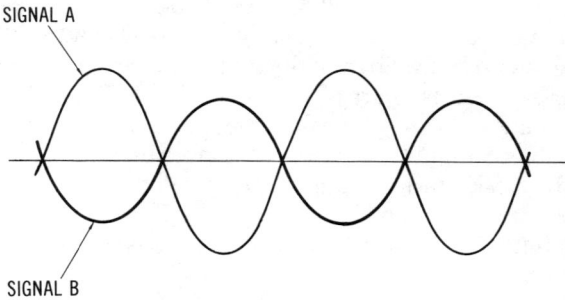

Fig. XIX-3. Signals 180° out of phase.

$$\text{Phase Shift} = \frac{\text{Time Shift}}{\text{Period}} \times 360°$$

For example, if the time shift is 70 microseconds and the period is 250 microseconds, the phase shift is $70/250 \times 360° = 100.8°$.

Occasionally, phase shift is expressed in radians. A radian is 57.3° ($360° = 2\pi$ radians). Expressing the phase shift in the example above in radians would give $70/250 \times 2\pi = 1.76$ radians.

Two separate measurements are required to determine the phase shift. In some computing counters, it is possible to set up the counter to make the two measurements and then automatically compute and display the phase shift directly. Special phase-shift measuring instruments have been developed to perform high-accuracy phase measurements. These devices are nothing more than dual-input counters optimized for phase measurements. They also include the special computing circuits that allow the period and phase-shift measurements to be made, the phase calculated, and the result displayed directly.

## COMPLEX SIGNAL MEASUREMENTS

While the majority of the signals whose frequency or time characteristics are to be measured are conventional and uncomplicated, there are other more complex signals that must often be measured. Some examples are rf bursts, pulse trains with different pulse widths and spacings, and signals with noise or distortion. Ordinarily, measuring such signals is difficult if not impossible for a conventional counter. However, some of the more advanced counters incorporate special circuits that make time and frequency measurements on complex signals fast and easy.

The counter circuits that facilitate such measurements are referred to as TI (time interval) *hold-off* or *arming* circuits. Their purpose is either to mask out an undesired part of the signal or to arm the counter to pass selectively a particular part of the signal to be measured.

The masking or arming signal is generated by an internal one-shot multivibrator whose pulse width is adjustable by means of a front-panel control. The mask or arming pulse is triggered by the input signal and is used in conjunction with the main-gate enable signal to control the internal gating. Some examples of armed or masked time and frequency measurements will best illustrate this feature.

**Frequency Measurements**

Fig. 6-11 shows several examples of complex waveforms whose frequency is to be measured. In Fig. 6-11A, the input signal is a repetitive rf burst. The objective is to measure the frequency of the rf in the burst, not the burst repetition frequency. To do this, an arming signal is set up. The beginning of each burst turns on the arming signal for a period of time set by the operator. An oscilloscope is used to adjust the arming duration so that it is equal to the burst time. With this arrangement, only the rf is applied to the counter main gate. It is assumed that the burst length is longer than the time-base gate duration to ensure a proper measurement.

In Fig. 6-11B, the input signal is the same rf burst as in Fig. 6-11A. Here a mask signal is used instead of an arming signal so that the burst repetition rate is measured instead of the burst frequency. At the

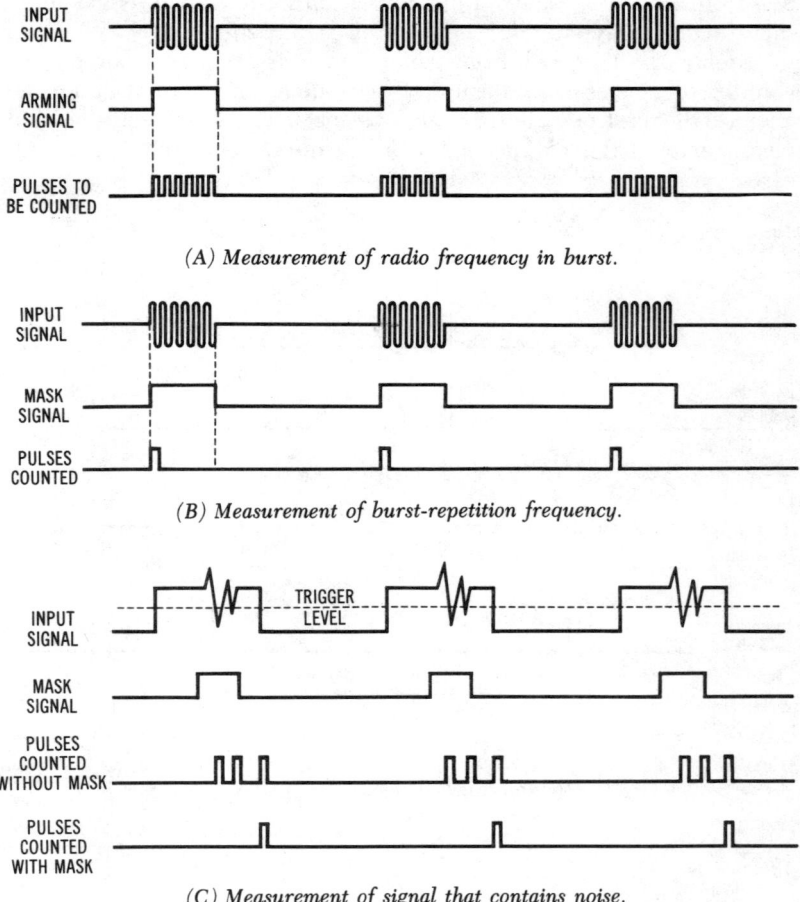

Fig. 6-11. Use of arming or masking capability for frequency measurements on complex waveforms.

beginning of each burst, the mask signal is generated, thus blanking out the burst cycles and preventing them from being passed through the main gate and counted. Instead, the counter simply counts the pulses generated at the beginning of each mask pulse. Therefore, the burst-repetition frequency is measured.

In Fig. 6-11C, we wish to measure the frequency of a signal containing noise "glitches." With the trigger level set as shown and a negative slope selected, the main gate will see the extra pulses generated by the noise. The result will be an incorrect frequency indication because the noise pulses will be counted. However, if a mask signal is used as shown, the noise pulses can be blanked out so that only the trailing edges of the input signal are counted.

A delay one-shot with front-panel control is often incorporated in a counter to give greater flexibility. The delay one-shot is connected ahead of the mask one-shot. The input signal triggers the delay one-shot, which later triggers the mask (or arming) one-shot. With this

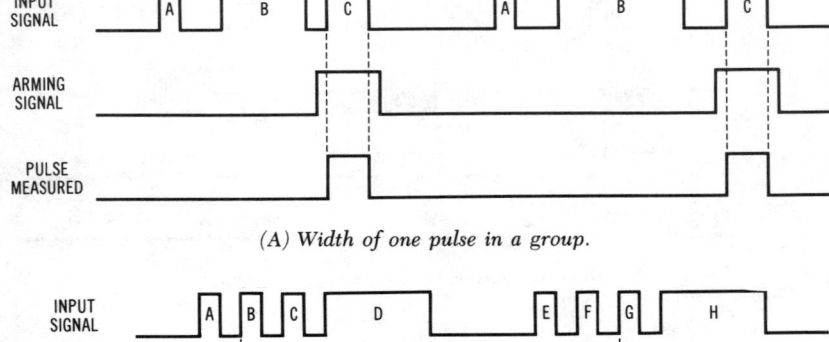

(A) *Width of one pulse in a group.*

(B) *Interval between selected pulses.*

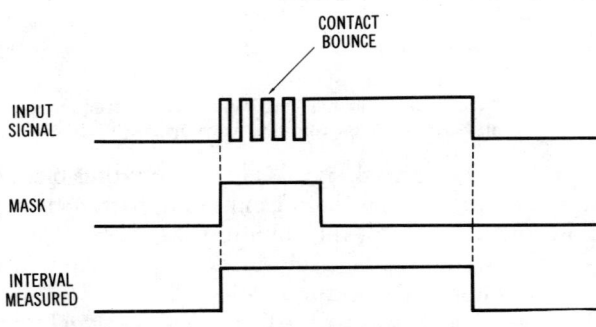

(C) *Elimination of contact bounce.*

Fig. 6-12. Use of arming and masking signals for time measurements on complex waveforms.

arrangement, the position of the mask pulse can be adjusted properly with respect to the input signal.

## Time Measurements

Occasionally it is necessary to measure some specific pulse characteristic in a complex pulse wave train. In such applications, an arming or masking capability is necessary. The examples in Fig. 6-12 show some typical measurements.

In Fig. 6-12A, we wish to measure the width of pulse C in the repetitive pulse train. The arming-signal delay and width are adjusted to pass only pulse C and mask out the remaining pulses. The counter is set up as usual to measure pulse width.

In Fig. 6-12B, we want to determine the time interval between the leading edges of pulses B and G. The arming signal allows only the leading edges of the desired pulses to be recognized. The intervening pulses are ignored by the counter.

In Fig. 6-12C, we wish to measure the duration that a push-button switch is depressed. When the switch is depressed, however, contact bounce occurs. This will confuse a counter set up to measure time interval. In fact, the counter may simply measure the time duration of the first contact bounce pulse and nothing else. To correct this problem, a mask signal is used. The mask allows the counter to ignore the bounce and simply measure the time interval during which the push button is closed.

CHAPTER 7

# Special Counters

Up to this point, the focus of this book has been on counter test instruments that use traditional counting circuits and techniques. We have emphasized common general-purpose counters because they are by far the most widely used. They are low in cost, versatile, and widely available. Most time- and frequency-measuring applications can be adequately fulfilled by these instruments.

However, there are many special counters that perform a variety of unique, unusual, or demanding functions. The purpose of this chapter is to introduce and explain some of these important special counters.

The subjects to be covered include counters that contain calculating and computing capability. These test instruments incorporate sophisticated computing circuits that are capable of performing a variety of mathematical and statistical calculations. Most of these computing counters employ a microprocessor. These counters fully automate time and frequency measurements and help to speed up and simplify the measurements while reducing errors.

Another category of special counters includes those used in industry. Industrial counters are a special class of counter used in monitoring and control applications. Both mechanical and electronic counters are used in a variety of manufacturing and process-control situations. These low-cost counters find application where time, frequency, events, and other characteristics must be measured and displayed.

## COMPUTING COUNTERS

The most sophisticated, expensive, and capable counters available today are the computing counters. These top-of-the-line instruments are called computing counters because they incorporate digital computing circuitry. When these capabilities are combined with the standard measurement circuitry of conventional counters, a whole new range of operations can be performed.

Modern computing counters incorporate a microprocessor. Early computing counters, first available a little over ten years ago, used MSI circuits to perform limited calculations. The most common calculation was division or finding the reciprocal so that the period of a signal could be measured and the frequency then computed and displayed. Of course, a microprocessor provides a much wider range of capabilities.

A computing or programmable counter contains two major sections, the measurement section and the computation/automation section. The measurement section consists of all of those conventional counting circuits described elsewhere in this book. These are the counters, registers, control circuits, time base, input-signal conditioning, and display circuits.

The computation and automation section generally contains a microprocessor. This small general-purpose computer is especially programmed to perform a variety of special measurement and computing functions. Its control program is stored in a ROM. Since the microprocessor is a real computer, it can manipulate the digital data accumulated by the measurement section in a variety of ways. Data accumulated in the measurement section can be transferred to the microprocessor and stored in memory. Whole sequences of measurements can be stored. The data can then be further modified or used in a variety of ways. It can be used in mathematical computations. Other numbers can be added to it or subtracted from it. The number can be mulitiplied or divided by a constant. Its reciprocal or square root can be found.

The microprocessor can also make decisions. Given the information accumulated by the measurement section, it can determine if the measurement data is above, below, or equal to a previously programmed level. Finally, the microprocessor can fully automate the measurement process. It can control trigger level and slope, time-base selection, mode, and other functions. It can be programmed to perform a sequence of measurements and computations. Programmable computing counters usually incorporate a keyboard similar to that used on calculators. This keyboard usually replaces all of the traditional front-panel controls ordinarily available on a standard counter.

Computing counters are one of a growing class of "intelligent" instruments. Intelligent instruments are those that incorporate microprocessors to automate the measurement process, provide improved accuracy, and simplify measurements. Other intelligent instruments being made today include digital voltmeters, oscilloscopes, function generators, and others.

## Functions of Computing Counters

Just what are some of the special operations that computing counters perform? One of the first things that a computing counter does when it is first turned on is provide a full self-test. The microprocessor accesses diagnostic routines stored in its ROM. In much less than a second, the microprocessor performs a complete check of all internal circuitry to assure the operator that the counter is operating properly and ready to perform. Some of the self-tests include a check of power-supply voltages, memory tests for the internal RAM and ROM, verification of the time base, and a complete exercise of the measurement section including the counters and display. A front-panel indicator light is generally used to indicate that the counter has been fully tested and is ready to go.

Another function usually performed by a computing counter is trigger-level setting and control. Most conventional counters provide a fixed trigger-level setting. Some advanced counters allow the trigger level to be adjusted by a potentiometer on the front panel. Voltmeters and oscilloscopes are often required to adjust the trigger levels properly. With a computing counter, trigger levels can be set digitally. This is done by entering the desired trigger level and polarity (slope) on the keyboard.

Automatic trigger-level setting can also be accomplished on some computing counters. When measuring signals whose input amplitude is unknown, the operator usually has to measure the signal with an external oscilloscope and then set the trigger-level adjustment. With a computing counter, the trigger-setting function can be fully automated. When a specific keyboard button is depressed the microprocessor proceeds to measure the incoming signal with an internal analog-to-digital converter. The positive and negative peak levels of the signal are determined. Then the counter computes the arithmetic mean of the two values. The trigger level is then automatically set to the mean value. On pulse signals, this means that the trigger level is set to the 50% amplitude point. At the mean or at the 50% amplitude level, trigger errors and false counting caused by noise and distortion are minimized or completely eliminated during most time and frequency measurements.

Another function of the modern computing counter is its ability to measure automatically pulse characteristics such as rise and fall time, width, duty cycle, and slew rate. With a conventional counter, characteristics such as rise time can be measured, but external instruments such as dvm's and oscilloscopes are needed. However, with a computing counter, these measurements can be made with a single instruction from the keyboard. For example, to measure rise time, the microprocessor automatically measures the positive and negative peak values of an incoming pulse. It then calculates the 10% and 90% voltage levels between which the rise time is measured. The trigger levels are automatically set at these points. A time-interval measurement is made and the results displayed. Fall time is measured in a similar manner.

Pulse characteristics such as width and duty cycle can also be measured with a computing counter. Pulse width is the time interval between the 50% amplitude points on the leading and trailing edges of the pulse. The computing counter measures the peak-to-peak amplitude of the pulse. The trigger levels of the two input channels monitoring the pulse are set to the 50% points, and the slopes are selected appropriately. The time interval (pulse width) is measured and displayed and/or stored. The counter can also measure the period of the pulse signal. If properly programmed, the counter can cause the pulse width to be divided by the period to compute the duty cycle, which can be displayed on the readout. These are only a few of the many automatic pulse measurements that can be performed with a computing counter.

The computing capability within the counter can be used for many different functions. The time or frequency value accumulated in the counter measurement section can be scaled, biased, or inverted to provide direct readout in other units. For example, frequency can be converted to revolutions per minute by multiplying the frequency by the appropriate constant. The period of an incoming signal can be measured with a counter and a frequency computed and displayed by a reciprocal operation. Bias or offset values can be entered via the keyboard and stored in the counter and thus added to or subtracted from a measured frequency value. For example, in a high-frequency radio receiver, it may be desirable to compute the received frequency by measuring the local-oscillator frequency and then adding or subtracting the fixed intermediate-frequency (if) value. The if value is stored in the counter along with the necessary mathematical operation. Usually, the local oscillator frequency, $f_o$, is higher than the received frequency, $f_r$, by the if value, $f_i$. Then $f_i = f_o - f_r$ or $f_r = f_o - f_i$. The local-oscillator frequency is measured; then the if value is automatically subtracted from it. The correct frequency of the received signal will be displayed. The microwave counter shown in Fig. 7-1 is also a computing counter and can perform this function. Because of the complete programmability of most computing counters, a great variety of computations and data manipulations can be made. This makes the computing counter one of the most versatile instruments available today.

Another common application would be the automatic measurement, computation, and display of phase shift. Most computing counters have two complete input sections and independent counters. To measure phase shift, the period of the incoming signals must be measured as well as the time shift between then. Once these measurements have been obtained, the phase shift between the two signals can be computed by

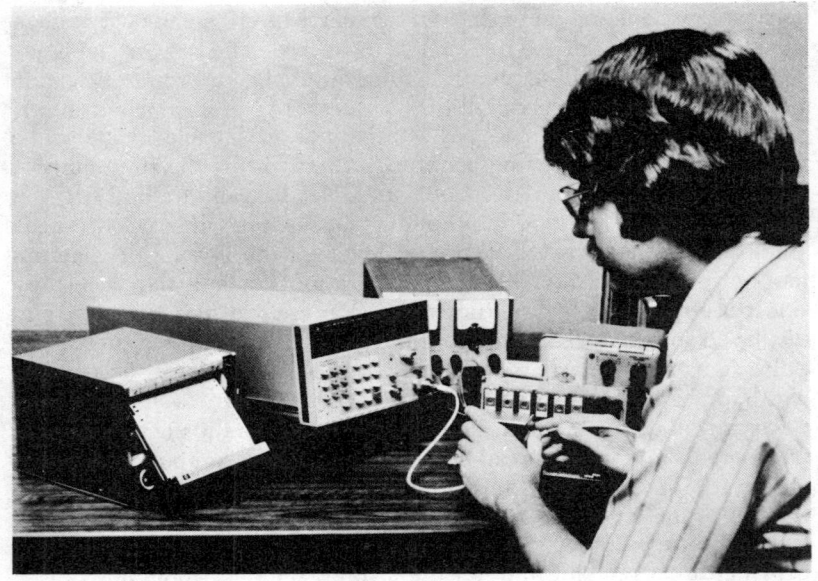

Courtesy Hewlett-Packard

Fig. 7-1. A microwave computing counter in use.

the counter. The time shift between the two signals is simply divided by the period and the quotient then multiplied by 360. The result is the phase shift in degrees, which is then displayed.

### Advantages and Benefits

The main benefit of a computing counter is its great flexibility. The computing counter allows more complex measurements to be made with less external accessory equipment. In addition, it allows these measurements to be made quickly, simply, and automatically with less error.

Measurement speed is one of the more important benefits of a computing counter. Because the operator does not have to set up a lot of front-panel controls or external equipment, the measurement from start to finish can be up to a thousand times faster.

Greater measurement accuracy is another advantage. Errors often introduced by the operator in setting controls improperly can reduce accuracy. The counter automates these functions and often selects the optimum settings for maximum resolution and minimum error.

Fewer external accessories are typically needed with a computing counter. In the applications chapter of this book, it is recommended that an oscilloscope always be used with a counter to ensure proper measurements. The scope allows you to look at the signal and, therefore, make the necessary front-panel adjustments to set the trigger level or perform other operations to minimize the effect of noise or signal distortion. With most computing counters, the scope and other external equipment such as a digital voltmeter can usually be eliminated. Despite this advantage, most individuals still like to look at the signal they are measuring. The use of an oscilloscope is still recommended even with a computing counter.

Finally, the benefit of more complex and sophisticated measurements is certainly an advantage. With standard counters, simple time and frequency measurements are relatively easy to make. But when these measurements are part of a more complex sequential testing process in a system, the benefits of a computing counter become obvious. These automated measurements greatly simplify the application.

## INDUSTRIAL COUNTERS

Industrial counters are a special class of digital counters used in a variety of industrial monitoring and control applications. These counters are not general-purpose test instruments. Instead, they are simply specialized devices that perform the single function of counting. They consist of either mechanical or electronic counting units and displays. They are small in size, low in cost, and designed to be dedicated to a single counting operation. Most industrial counters are designed for only very-low-speed counting operations.

Industrial counters are used in a variety of ways. Because of their small size and low cost, they can be readily built into laboratory equipment, electronic instruments, office machines, and other products. For example, they can be used in vending machines to count the number of items dispensed. They are often used in office copiers to keep track of the number of copies made. You will find them on a variety of data-processing equipment.

Industrial counters are also widely used in factories. They are used on assembly lines to count items being manufactured and are attached to all types of machinery to keep track of the number of operations performed. They are connected to various types of machine tools and devices such as printing presses. You will see them built into punch presses, conveyer belts, and many other production machines. In any application requiring the counting of packages, units, pieces, parts, operations, or even people, an industrial counter may be used. Industrial counters can also keep track of time, monitor quantity, measure values, total or tally events, and otherwise perform operations involving the counting of discrete actions.

Based on a predetermined count, industrial counters can be used to control external functions. Some of the more modern electronic industrial counters include computing or calculating capability as well.

In this section, we discuss the two most popular classes of industrial counters, mechanical and electronic.

**Mechanical Counters**

Mechanical counters are digital counting devices that use mechanical techniques to do the counting. Mechanical counters incorporate gears, ratchets, escapements, and other mechanical arrangements to count external events and display them. Typically, there is a counting mechanism for each of the digit positions in the counter. Many of the mechanism parts are made of plastic for light weight and high counting speed.

The mechanical counter can be actuated in a number of different ways. It can be actuated by purely mechanical techniques. For example, some mechanical counters can be manually operated by a push button. Others may be actuated by an external push or pull stroke. In some applications, the count mechanism is actuated by a rotating shaft which can include a gear train. A lever-operated mechanical counter is shown in Fig. 7-2.

Fig. 7-2. A lever-operated mechanical counter.

Courtesy Redington Counters, Inc.

Another, more widely used, method of actuating a mechanical counter is through an externally applied electrical signal. Such counters are more accurately referred to as *electromechanical counters*. A pulse of dc or ac is applied to the counter. This pulse actuates a solenoid to cause the mechanism to be stepped or incremented. Fig. 7-3 shows an electromechanical counter.

There are three basic methods of applying a dc or ac pulse to an electromechanical counter. See Fig. 7-4. In the first case, Fig. 7-4A, the ac power-line voltage is applied to a step-down transformer and then connected to the counter solenoid by an external switch. This can be a manually operated switch (such as a push button), a limit switch on a machine, or the contacts of a relay. In Fig. 7-4B, a dc voltage is applied to the counter solenoid. Again, some form of external switch or relay must be used to connect the dc voltage to the counting mechanism. The dc voltage in this case is derived from a small low-voltage power supply. A

Fig. 7-3. An electromechanical industrial counter.

Courtesy Redington Counters, Inc.

transistor switch can also be used to apply the pulse of voltage to the counter (Fig. 7-4C). This transistor is usually driven by other circuitry.

As indicated earlier, mechanical counters are essentially low-speed devices. They are typically rated in terms of counts per minute (cpm), counts per second (cps), or impulses per second (ips). The maximum counting speed for most mechanical and electromechanical counters is approximately 60 counts per second. This is best achieved with pulses. The counting rate with an ac signal is typically about half the maximum obtainable with dc input pulses.

In order to count reliably, an electromechanical counter must be supplied with voltage pulses of the correct duration. Because of the characteristics of the internal counting mechanism, the electrical pulse must be applied for a certain minimum duration and then terminated. A pulse of less than the required duration will not step the counter. However, if the pulse is longer than necessary, it could limit the upper counting rate.

When pulses are applied to the solenoid through a transistor, care must be taken to protect the transistor. High-voltage transients are generated when the transistor turns off. This is similar to the effect obtained when the voltage is removed from a relay coil. The collapsing magnetic field around the coil induces a spike of very high voltage that can in some cases damage the external switch or electronic circuitry associated with the coil. Various types of spike-suppression techniques are used to eliminate this problem. The simplest form of suppression is simply to connect a diode across the coil as shown in Fig. 7-4C. When the pulse of voltage is removed from the coil, the collapsing magnetic field induces a high voltage in the coil. The polarity of the voltage will be such that it will cause the suppression diode to conduct, and, therefore, the voltage spike will be clamped to a very low level, thus protecting the transistor. The problem with this suppression technique is that the induced voltage causes the diode to conduct and allows current to continue to flow in the coil. For this reason, the coil remains actuated for a longer period of time. This suppression technique can often slow down the counting rate. Other types of spike-suppression circuitry involving a combination of capacitors, resistors, and diodes can also be used to

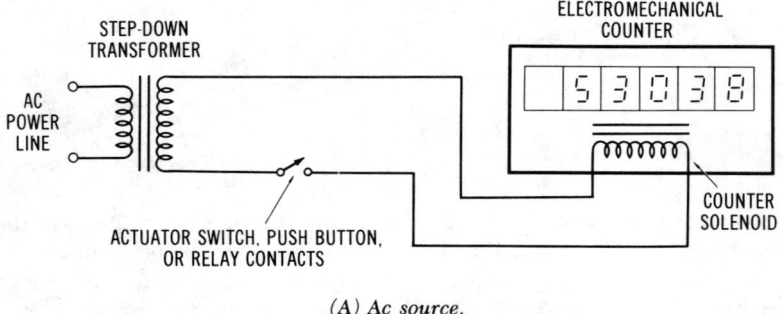

*(A) Ac source.*

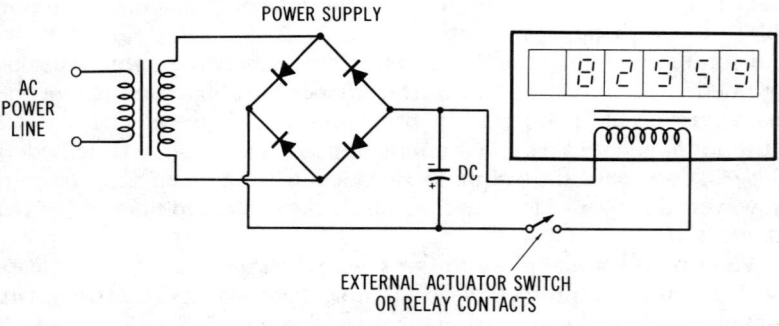

*(B) Dc source.*

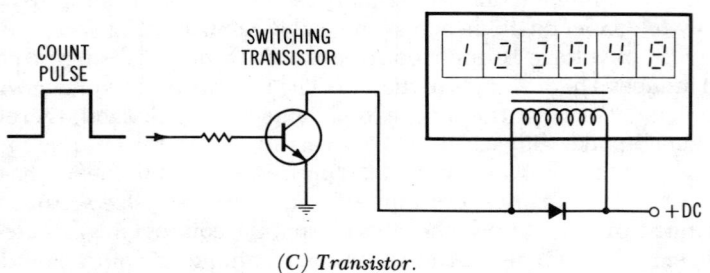

*(C) Transistor.*

Fig. 7-4. Methods of driving an electromechanical counter.

minimize damage caused by induced voltage spikes but at the same time allow the coil to be de-energized more quickly

Electromechanical counters can be obtained in a wide variety of forms. Coil-voltage ratings of 6, 12, 24, 48, 110, and 220 volts in both dc and ac versions are standard, but custom values can also be obtained. Different maximum count capabilities are also available. The simplest is the single-digit counter that counts from 0 through 9 or 1 through 10. Counters with any number of digits between 1 and 8 can be readily obtained.

A special class of mechanical counter is the pneumatic counter. Like mechanically and electrically operated counters, these devices feature a mechanical counting and display system. However, the count mechanism is actuated by pneumatic techniques. A high-pressure blast of air from some pneumatic source is required to increment or decrement the counter.

There are several different classes of mechanical counters used in industrial control applications. Some of the more popular types in use are described below.

*Stroke Counters*—Stroke counters feature mechanical counting and display mechanisms. The actuation of the mechanism is through some physical linkage. In the smaller mechanical counters, the actuation is by finger on a button or by hand on an external lever. Larger mechanical counters are used where the mechanical operations of machines must be tallied. In many applications, the machine to which the counter is attached has a moving part which can be attached by a lever spring or cam to the counter actuator arm.

*Electrical Totalizing Counters*—The electrically actuated mechanical counter is by far the most popular and widely used type of mechanical counter. Its primary function is simply to tally, or totalize, the occurrence of external events. A switch, relay, or other sensor is used to detect the operation or event and actuate the counter.

Electrical totalizers are available with either a mechanical or electrical reset feature. Reset capability on a counter refers to the ability to set the counter to zero. At the beginning of a count operation, it is usually necessary to zero the counter so that the digits will indicate the proper number of event occurrences. In many counters, the reset operation is simply accomplished with a front-panel push button. In other counters, an electrical reset capability exists. To reset such a counter, an external dc or ac pulse is applied to a separate solenoid that causes all counter digits to be reset to zero.

*Lineal (Measuring) Counters*—Another type of counter is used to make distance or length measurements. The counter and display are mechanical in nature and are actuated through a rotating shaft. This rotating shaft is externally connected to some type of measuring wheel, the circumference of which is carefully selected to perform the correct measurement. This wheel rolls along the linear surface to be measured. This could be a bolt of cloth, a roll of paper, or a sheet of steel. Wire, thread, or similar items can also be measured by this technique. As the wheel rolls along the item to be measured, the counter mechanism is stepped or incremented at the appropriate times and displays the measurement in inches, meters, or some other linear units. It is the size of the wheel and the nature of the gearing that determines the units to be displayed.

*Predetermining Counters*—A predetermining counter is one that allows a specific count value to be preset. When the counter reaches the preset value, it normally stops counting. In most preset counters, some external control operation is initiated when the predetermined value is reached. For example, in most electromechanical predetermining counters, when the count value is reached, a relay is actuated. The relay contacts are made available to control external devices. For this reason, predetermining counters are control counters. They can be used to operate signal lights, ring bells, start or stop motors, or otherwise control operations. For example, predetermining counters can provide production control because they are useful in controlling losses due to overruns or underruns. They give precise counting and control in a variety of applications.

There are two basic types of predetermining counters. In one type, the preset value is manually entered with some form of push-button mechanism. The counter is then reset to zero before the counting begins. The counter is incremented with each input event. When the predetermined value is reached, the internal relay contacts are actuated. Either a contact opening or a contact closing may be obtained for starting or stopping the external operation.

In the other type of predetermining counter, the preset value is set into the counter manually as before. However, the counter is not reset. Instead, it is decremented from the preset value. In other words, the counter counts down from the preset value. When zero is reached, the internal relay contacts are actuated, and the count operation ceases.

*Up/Down (Add/Subtract) Counters*—Most mechanical totalizing counters are up counters in that each time an external event occurs, the count is incremented. However, some decrement counters are also available. A special class of counter is that which can count up or count down. In other words, it can add or subtract external counts. In these counters, the counting unit is also mechanical, but the mechanism can be driven in two directions so that counts can be added or subtracted. This is done with two separate input mechanisms and solenoids. Pulses to increment the counter are applied to one set of input terminals. Another set of input terminals is provided for the decrement pulses. These units are also referred to as *bidirectional counters* by some manufacturers.

*Printing Counters*—A printing counter is a mechanical counter that also contains a printing mechanism. This mechanism causes the numbers displayed by the counter to be printed on a thin strip of paper similar to that used in adding machines. Various printing mechanisms are used, but by far the most popular is the impact printer, which uses an inked ribbon and a wheel with raised numbers like those on a typewriter arm. An external electrical signal causes the print operation to take place. Printing counters allow a permanent record to be made of the various counting operations performed.

*Advantages of Mechanical Counters*—With electronic counters so low in cost and with so many LSI counting circuits available, just why are mechanical and electromechanical counters so widely used? Mechanical counters offer numerous advantages over their electronic equivalents.

The main advantage of a mechanical counter over an electronic counter is its ability to withstand hostile environments. Mechanical counters can tolerate higher temperatures and even nuclear surroundings. They are much more tolerant of noise and severe power-line transients that could damage electronic counters or cause false counting. Mechanical counters are capable of withstanding strong electromagnetic interference fields generated by surrounding equipment. Mechanical counters are also more rugged. Factories and processing plants typically have extremely hostile conditions in which even the best of electronic counters may fail.

Another advantage of the mechanical counter is its nonvolatility. Volatility refers to the inability of a counter to hold its count value when electrical power is removed. In an electronic counter, data is usually stored in flip-flops. If power is removed from them, the data is lost. This will not occur in mechanical counters because the mechanism automatically stores the count value.

Mechanical counters are often lower in cost for a given application than an electronic counter. Electronic counters tend to be more sophisticated and thus are more expensive. Mechanical counters, on the other hand, are simple, straightforward, and usually low in cost. For simple applications, a mechanical counter is by far the best solution. Electronic counters are often expensive overkill.

## Electronic Counters

While mechanical and electromechanical counters have predominated in industrial control applications, electronic counters are quickly replacing them. In addition, because of their increased capabilities, electronic counters are opening up new applications for counting techniques in various industrial monitoring and control situations.

Electronic counters perform virtually the same functions as mechanical and electromechanical counters. Electronic counters for totalizing, predetermining, up/down counting, linear measurement, and printing are available. An electronic predetermining counter is shown in Fig. 7-5. It uses LED displays and thumbwheel switches to enter the preset value.

While it has been possible to build electronic industrial control counters for many decades, their cost has always been significantly more than the cost of equivalent mechanical or electromechanical units. In low-speed, dedicated applications, mechanical counters meet the need at the lowest possible cost. But with developing semiconductor technology, the price of electronic counting circuits has dropped considerably year after year. With the introduction of MSI integrated circuits, the cost of electronic counting circuits became practical for many industrial applications, particularly those requiring features that needed the advantages of electronic circuitry.

Today LSI integrated counting circuits are available. The entire counter can be readily placed on a single silicon chip. As a result, extremely sophisticated electronic counters can be made at very low cost. In addition, low-cost LED and liquid-crystal displays are available to implement the digital readout.

Today, electronic counters are lower in cost than ever, and in many cases they are competing with mechanical and electromechanical counters in industrial applications. The typical modern electronic counter uses an LSI chip with either LED or liquid crystal (LCD) readouts. The size and cost of such sophisticated counters are equivalent to those of their mechanical counterparts.

What are the advantages of an electronic counter over its mechanical equivalent? The main advantage of an electronic counter is its ability to count at higher speeds. As indicated earlier, the upper count limit in a typical mechanical or electromechanical counter is approximately 60

Courtesy Industrial Timer Corp.

Fig. 7-5. An electronic predetermining counter.

counts per second. While this is adequate for many industrial applications, higher speeds are often necessary. As you know, electronic counters are capable of extremely high count rates. However, in most industrial applications, speeds above several thousand counts per second are rare. Most electronic industrial counters are designed to count up to approximately 10 kHz. Few applications require anything beyond this.

While high counting speed is the main advantage of electronic counters, such counters are also often used in low-speed counting applications. The reason for this is that an electronic counter can recognize and count very narrow input pulses. This is not always possible with an electromechanical counter. In order for an electromechanical counter to count reliably, the input must generally have a minimum specified duration. In most electromechanical counters, this is in the range from 20 to 60 milliseconds. Narrower pulses simply will not be recognized or counted. Electronic counters, however, can readily accept pulses as short as several microseconds. Even though the counting rate may be no more than 30 pulses per second, if the pulses are narrow, an electronic counter is the best solution for this counting problem.

One benefit of an electronic counter over an electromechanical unit is its higher reliability. While mechanical counters can count many millions of times before they fail, electronic counters have an almost unlimited count life. Integrated circuits and other solid-state components are extremely reliable and for long-term counting applications will provide the best reliability.

Finally, one of the greatest advantages of an electronic counter is its versatility. There are many more things that can be done with electronic circuitry than with mechanical components. Many sophisticated operations can be performed. For example, electronic counters have the ability to measure time. Because of their higher count-rate capability, electronic counters for measuring time intervals (Fig. 7-6) can be designed at very low cost. Industrial timers can be used to measure the time between the occurrence of two external events or to initiate some action after a certain length of time has elapsed. A special version known as *digital clock* (Fig. 7-7) can initiate operations at preprogrammed times. Many other timing applications are possible.

Another example is computation. Just as there are special computing counter test instruments, so are there computing industrial counters. These units are usually programmable to perform one of several computing functions. Multiplication and division are typical functions.

Computing industrial counters usually accept an input from an external transducer. This could be a switch or a photocell, but most applications use some form of magnetic transducer to count gear teeth or some similar measure of rotation. For example, see Fig. 7-8. Here, a

Courtesy Electronic Research Co.

Fig. 7-6. An industrial electronic timer.

Courtesy Electronic Research Co.

Fig. 7-7. A digital clock for use in industrial control applications.

gear with 100 teeth is mounted on a motor shaft. A magnetic transducer adjacent to the teeth generates a pulse for each tooth. These pulses are counted by the counter. Since there are 100 teeth, there will be 100 pulses generated per revolution. By counting the pulses for a known period of time, the speed can be determined and displayed. If the time interval is 600 milliseconds, the count will be read out in revolutions per minute (rpm). Assume that the motor speed is 2000 rpm and that a 100-tooth gear transducer is used. This translates to $100 \times 2000 = 200{,}000$ transducer pulses in one minute, or $200{,}000/60 = 3333.33$ pulses per second. During a gate time of 600 milliseconds, or 0.6 second, $0.6 \times 3333.33 = 2000$ pulses will be counted and displayed. In

170

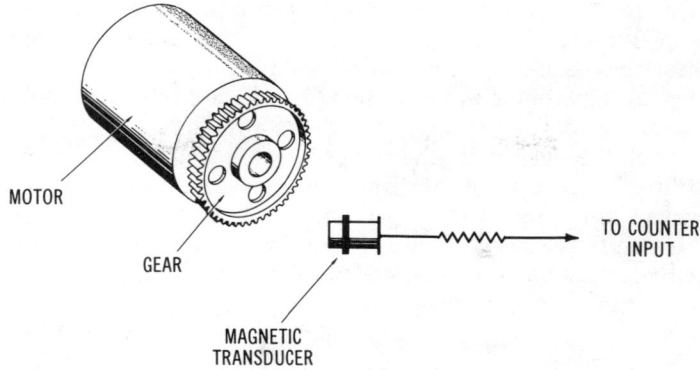

Fig. 7-8. Use of magnetic transducer to generate pulses from gear teeth to measure motor speed.

some counters, the gate time can be changed so that any desired units may be displayed. In other counters, the transducer pulses can be multiplied or divided by the counter so that virtually any unit may be displayed (revolutions per second, etc.).

Depending on the type of transducer used, almost any kind of measurement can be made with a computing counter. The rate of flow of liquid in a pipe can be determined with an appropriate transducer. In Fig. 7-9, for example, the magnetic transducer generates pulses from the rotary vane, which turns at a speed proportional to the rate of flow of the liquid. The flow rate can be measured and computations made to display it in gallons per hour, liters per minute, or pints per second.

With a linear measurement wheel, the rate of speed of a roll of paper or other material can be measured and displayed. The computing counter makes it easy to match the transducer to the counter and display the measurement in virtually any units.

What are the disadvantages of electronic counters? One is their

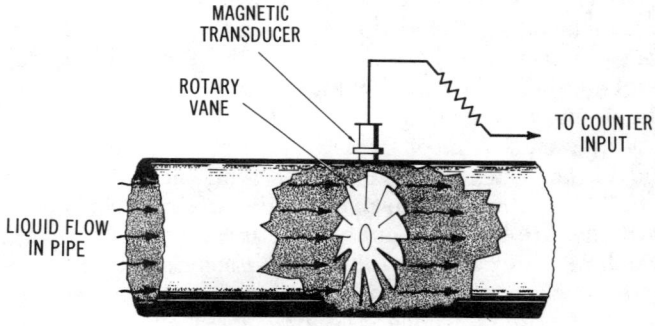

Fig. 7-9. Use of magnetic transducer with rotary vane to measure liquid flow in a pipe.

higher cost. As costs continue to drop, electronic counters will no doubt take over most of those applications formerly held by mechanical counters. Some electronic counters have been ruggedized to withstand the severe environmental conditions to which industrial counters are often subjected.

One of the major disadvantages of an electronic counter is its volatility. When a power failure occurs, the count information is lost. However, this disadvantage has been overcome in some electronic counter units. This is done by installing an internal battery to operate the circuitry should a power-line failure occur. The battery is not used under ordinary operating conditions, but when power is removed, the battery automatically cuts in and retains the count information. Since most electronic counters have a significant current drain, battery life is short, but since the battery is used only during a brief ac power failure, battery operating time is usually negligible. Some electronic counters have been implemented with CMOS circuitry and liquid-crystal displays. Both have extremely low power consumption, and as a result, battery-only operation is practical.

**Sensors**

All industrial counters require an input signal from some external device to cause a count to occur. This input signal usually comes from a sensor or transducer. A sensor is simply a device that detects the presence or absence of an item to be counted. A transducer, on the other hand, typically converts some physical motion into a pulse that can be counted by the counter. A wide variety of sensors and transducers are used with counters in industrial applications. A summary of the most popular types is given here.

*Switches*—The most widely used sensor is also the simplest. It is nothing more than a switch with two electrical contacts that can be opened or closed to generate a signal for the counter. The switch is usually connected in series with a dc or ac power source and the counter. Usually, when the switch is closed a pulse will be generated. The switch can be a manually operated device; each time an event occurs, an operator can actuate the switch, thus recording the count. Switches can also be set up so that they are operated by the items to be counted. For example, items passing along a conveyer belt can be used to actuate a switch as they pass. A mechanical arm attached to the switch can be operated so that when an item hits the arm, the switch is closed to generate a pulse. Limit switches can also be used to actuate the counter. Limit switches are usually mounted at critical points on a machine to signal the limits or extremes of operation. Each time a machine stamps out a part or performs some other operation, the movement of the machine can be used to actuate the limit switch, thereby generating a count pulse.

*Relays*—A relay is an electrically operated switch. A coil forms an electromagnet that operates a set of switch contacts. The relay is a common way of generating a pulse for an industrial counter. This is done by connecting the relay coil to some electrical source. Each time current is applied to the relay coil, the contacts close, and a pulse will be generated and recorded by the counter. Relays are useful in those applications in which an electrical signal applied to a machine or some other device can be used to operate the relay coil.

*Reed Switches*—A reed switch is a pair of contacts enclosed in a tiny glass tube. The contacts are actuated by the application of an external magnetic field. A coil wound over the glass tube can be used to create a reed relay; the contacts close (or open) when the coil is energized and a magnetic field is produced. However, reed switches are most often actuated by permanent magnets that are passed near them. When the field of the permanent magnet comes near the reed switch, the contacts close. Typically, the reed switch is permanently mounted in position, and the device to be counted is supplied with a permanent magnet.

*Photocells*—Next to switches, photocells are by far the sensors most widely used with industrial counters. A photocell is a light-sensitive device that generates a voltage or undergoes a change in electrical resistance when light is applied to it. Most photocell setups consist of a separate light source and a photocell unit. The items to be counted are passed between the photocell and the light source. With the light shining on the photocell, no action takes place, but when the item to be counted passes between the two, the light to the photocell is shut off, and a count pulse is generated.

Another type of photocell unit widely used in industrial applications is the reflective photocell. This is a device that contains both the light source and the photocell in a single housing. The light source directs a light beam outward toward the item to be counted. Typically, the item will be an excellent reflector of light; if it is not, a reflective surface of some type can be attached. Light striking the reflective surface is reflected back into the photocell, and a pulse for the counter is generated.

Miniature photodetector units have been developed for very tiny objects. These often include a light-emitting diode as the light source and an infrared photodetector.

In most cases, the signal generated by a photocell cannot be used to trigger the counter directly. Normally, the photocell output is amplified and shaped by a Schmitt trigger before being counted.

*Proximity Switches*—A proximity detector is a device that senses the presence of an object and generates an appropriate output pulse. Most proximity detectors are designed to sense metallic objects. One type of proximity detector is the magnetic sensor. This device consists of a permanent magnet with a coil wound around it. The unit is pointed toward a metallic object, which must be of a ferromagnetic material such as iron or steel. When the iron or steel piece passes in front of the unit, the magnetic field generated by the permanent magnet is disturbed, and a small voltage is induced into the coil. This pulse is then amplified and shaped to generate an appropriate input for the counter.

Another type of proximity sensor is the rf sensor. It consists of a coil which is part of the resonant circuit in a transistor oscillator. When a metallic object enters the magnetic field produced by the coil, the metallic object absorbs energy from the coil. This causes the oscillator to stop oscillating or reduces the amplitude of its output. The adjacent metal can also cause a frequency change. Any of these conditions can be sensed by other circuitry to generate an output pulse for a counter.

*Rotary Pulse Generators*—A rotary pulse generator, also sometimes referred to as a *shaft encoder*, is a device that generates a fixed number of equally spaced pulses for each revolution of the shaft. Some of these pulse generators look very much like an electric motor in a cylindrical housing with a single input shaft. This shaft is coupled to a motor or other rotary device.

Internally, the pulse generator may consist of a gear and a magnetic pickup, or it may contain a disk with closely spaced alternately clear and opaque areas that pass between a light source and a photodetector. In either case, as the shaft is turned, pulses are generated. The number of pulses generated per revolution depends on the unit and can vary from one pulse per revolution to as many as several hundred per revolution. When applied to a counter, these pulses can be used to measure shaft rotational speeds, or they can be used for making linear measurements.

## Applications

The uses of industrial counters are best illustrated by actual examples. Several practical applications are shown here.

Fig. 7-10 shows how items on a production line can be counted. As the items are manufactured, they are placed on a conveyer belt, counted, and transferred to an area where they are packed into boxes. A light source and photocell are used to sense the items and generate pulses for the counter. A manual reset button on the counter is depressed so that the count starts at zero. The conveyer belt then begins to move as the manufactured items are placed on it. As each item passes between the light source and photocell, it is tallied on the counter.

There are some special conditions that may make this rather common application more sophisticated. For example, suppose that we want to count by dozens instead of by single items. A computing counter would be ideal in this application. It would effectively divide the input by 12 and increment the counter once for every 12 input pulses. The display would read dozens.

Suppose further that problems sometimes occur on the conveyer belt.

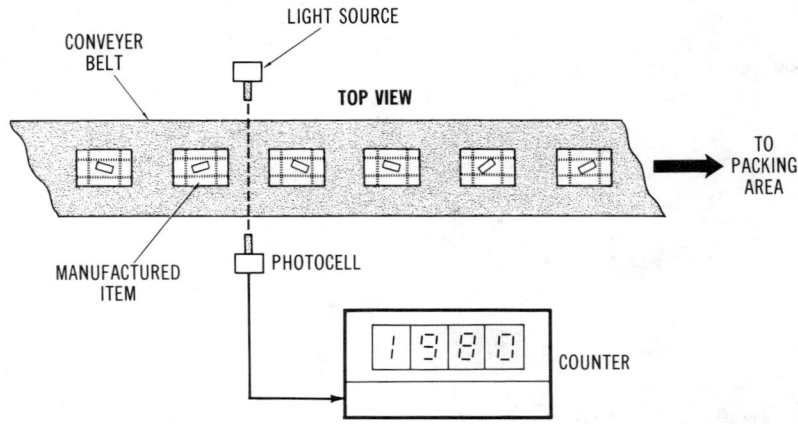

Fig. 7-10. Method of counting items on a moving conveyer belt.

An item is placed on the belt incorrectly and falls off or jams the unit, causing it to halt. The conveyer belt has to be reversed for several feet to clear the problem. Reversing the direction of the conveyer belt will cause previously counted items to be counted again, thus creating a false count. The photocell and counter cannot distinguish one direction from the other.

In this application, an up/down counter could be used to keep an accurate count. The counter is incremented (counts up) as long as the belt moves in the right direction, but if the motion is reversed, the counter is made to count down, thereby subtracting counts. When the proper direction is resumed, the up count proceeds normally. With an up/down counter, the correct count can always be displayed. Because of the simplicity and low speed of this application, a low-cost mechanical unit would be the best choice.

Another industrial-counter application is illustrated in Fig. 7-11. Here, a long strip of paper from a supply roll is being pulled along by a pair of drive rollers. The objective is to cut the paper into desired lengths. A predetermining counter is used to perform this function.

The measuring operation is performed by the upper roller, the rotary shaft transducer, and the counter. The roller has a *circumference* of 18 inches; therefore, for each revolution, 18 inches of paper passes the roller. A rotary transducer attached to the roller generates a fixed number of pulses per revolution. For example, if the transducer produces 180 pulses per revolution, then 180 pulses are generated per 18 inches. This translates to 10 pulses per inch. One pulse then represents 0.1 inch. The pulses are counted by the predetermining counter. With this arrangement, the paper can be accurately measured to the nearest 0.1 inch.

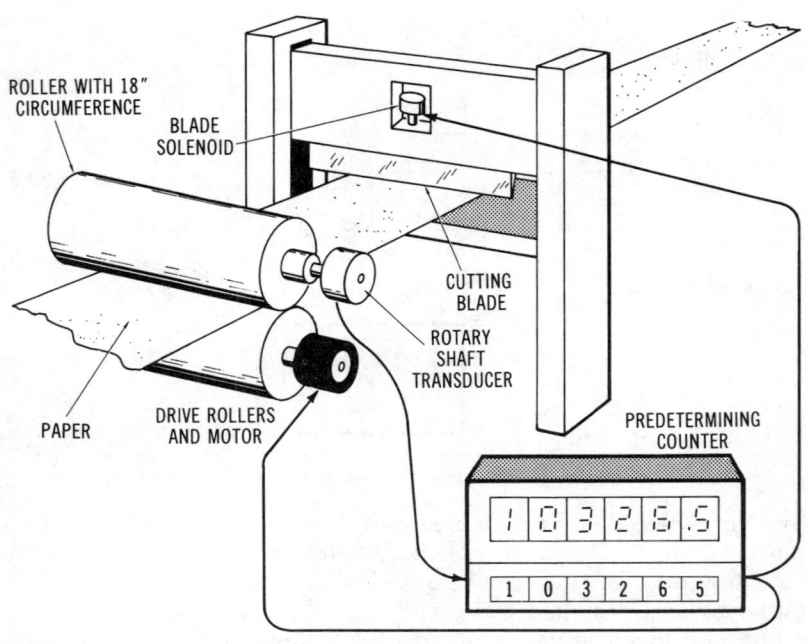

Fig. 7-11. Predetermining counter used in cut-to-length operation.

Cutting the paper to a specific length is accomplished by presetting the counter to the desired length. Suppose the desired length translates to a 10,326.5 preset value. This value is entered into the counter via thumbwheel switches or by other means. The counter is first reset to zero. Then the paper begins to roll. The counter accumulates pulses as the paper rolls by. When the predetermined count is reached, the counter outputs control signals that stop the drive motor and actuate a solenoid that drives the cutting blade to cut the paper. The cycle is then repeated. Either a mechanical or electronic counter could be used in this application, depending on the speed of the rollers.

The use of an electronic timer is illustrated in Fig. 7-12. In this example, the timer is used to measure the speed of cars passing over a given section of road. The key to this measurement is the basic formula

$$\text{Rate} = \frac{\text{Distance}}{\text{Time}}$$

The rate, or speed, is a function of the time it takes the car to traverse a known distance. To measure speed, then, we mark off a fixed distance on the road and measure the time it takes for the car to travel that distance.

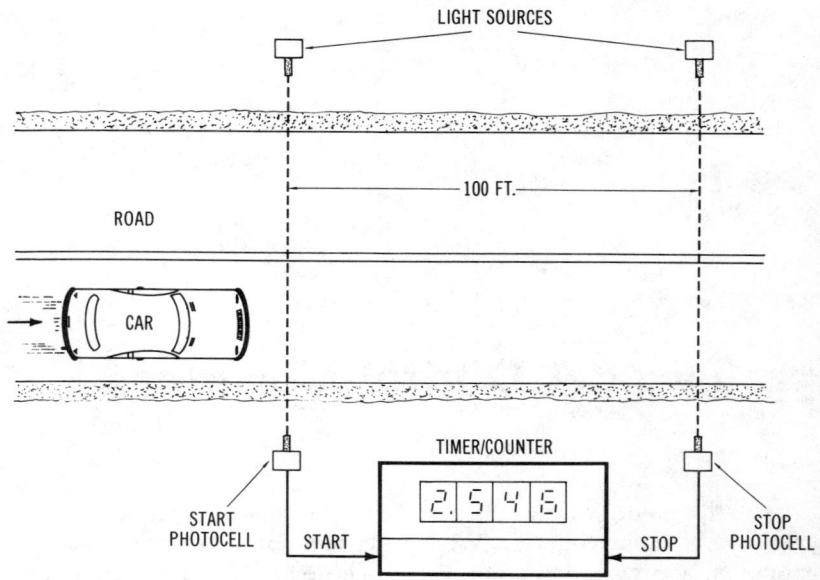

Fig. 7-12. Timer used to measure speed of car.

Assume the known distance is 100 feet. We place a light source and photocell at the beginning and end of the 100-foot section. These photocells are used to start and stop an electronic timer. The counter in the timer is allowed to accumulate 1-kHz pulses while the car travels the known distance. With 1-kHz pulses, the time measurement resolution is 1/1000 = 0.001 second, or one millisecond.

To begin a measurement, the counter is first reset to zero. The car then drives toward the marked section of road. When the car breaks the light beam of the first photocell, the timer begins counting 1-kHz pulses. When the car passes the second photocell, the count is stopped. The counter displays the time. Suppose the measured time for 100 feet is 2.546 seconds. The speed then is 100/2.546 = 39.277 feet per second. Usually, this must be converted into miles per hour (mph). A mile is 5280 feet and one hour is 3600 seconds. A speed of one mph is 5280/3600 = 1.467 feet per second. The car speed then is 39.277/1.46 = 26.78 mph. This computation can be made with a calculator after the measurement is made. If a computing counter is used, it can be programmed to make the time measurement and then compute and display the speed directly in miles per hour.

CHAPTER 8

# Typical Digital Counters

Up to this point, we have discussed only the general organization and operation of digital counters. We have concentrated on their modes of operation, accuracy, and applications. These are the more important topics as they apply directly to your own selection and use of a digital counter. It is not really necessary to know the circuit details of a digital counter to use it properly. However, some familiarization with the circuitry will give you a better understanding and appreciation of the operation, capabilities, and limitations of a counter. It may even improve your ability to apply the counter and interpret its results.

This chapter contains descriptions of several typical digital counters. The organization and operation of these counters is discussed. Where appropriate, details of the electronic circuitry are also considered.

Four modern electronic counters will be considered in this chapter. The first is a low-cost, handheld, battery-operated integrated circuit unit. Its upper frequency limit is not high, but it is a satisfactory counter for everyday noncritical frequency measurements. This counter illustrates the impact of large-scale integrated circuits on counter design, size, and cost.

Another hand-held, battery-operated counter is considered next. It, too, uses LSI circuits. This counter is designed for portable measurement of radio frequencies to 512 MHz. Its main use is frequency measurement in mobile radios or handheld transceivers.

A more conventional bench counter is considered next. This is a general-purpose test instrument implemented with MSI integrated circuitry. It covers frequencies of up to 225 MHz. This is typical of many units in use today.

Finally, a universal timer/counter is considered. This high-frequency unit is capable of a variety of frequency and time measurements. It uses a combination of MSI and LSI circuits to implement the various functions. A microprocessor-based interface is used for connecting the unit into a test system.

## HANDHELD LSI COUNTER

The counter to be analyzed in this section is the Conar Model 202 (Fig. 8-1). This is a low-cost kit frequency counter designed for general-purpose frequency measurements. Such a counter is widely used by electronic service technicians, hobbyists, and experimenters. This counter is also supplied as part of the NRI Schools home-study courses in communications and electronic servicing.

**Specifications**

The specifications for this counter are listed in Chart 8-1. This instrument features a six-digit LED display of the type used in handheld electronic calculators. The decimal point is permanently fixed between the second and third digits from the right. This means that there are four

Courtesy NRI Schools

Fig. 8-1. Conar Model 202 LSI counter.

## Chart 8-1. Specifications of Conar Model 202

| | |
|---|---|
| Frequency Range: | 10 Hz to 3 MHz, minimum |
| | Maximum frequency range typically 6 MHz |
| Number of Display Digits: | 6 |
| Resolution: | ±10 Hz, ±1 count |
| Input Impedance: | 50,000 ohms shunted by 20 pF plus cable capacitance |
| Sensitivity: | 100 mV rms maximum |
| | 50 mV rms typical |
| Maximum Input: | 200 volts, ac and dc combined |
| Power: | 6 volts, four AA penlight cells |

digits to the left of the decimal point and two to the right. The display reads the frequency in kilohertz; therefore, the two least significant digits, those to the right of the decimal point, read a fractional part of a kilohertz. (These two digits indicate hundreds of hertz and tens of hertz, respectively, from left to right.) For example, if the display shows 3542.01, the frequency is three thousand five hundred forty-two point zero one kilohertz. This can also be expressed in megahertz by simply moving the decimal point three digits to the left, or 3.54201 MHz. The least significant digit represents 10 Hz; therefore, the resolution is 10 Hz. Since the decimal point in the display is fixed, there are no other frequency ranges. In other words, there is a single fixed time base. No accuracy figures are given for this counter. The time base is derived from a 6.5536-MHz crystal.

The input sensitivity is typically 50 millivolts. This means that it takes at least 0.05 volt to trigger the unit reliably. The sensitivity can be as high as 100 millivolts on some units. These are rms and not peak or peak-to-peak values.

The input impedance of the counter is 50,000 ohms in parallel with a shunt capacitance of approximately 20 pF. The total input capacitance is 100 pF if the cable supplied with the unit is used. A 50,000-ohm input impedance is low compared to that of most typical counters. However, it is sufficiently high to minimize the loading on most circuits. Most integrated circuits, both digital and linear, have relatively low output impedances and will not be affected by the 50,000-ohm impedance. Should loading occur, in most cases it will not affect the frequency measurement. But, be careful in measuring oscillator circuits; excessive loading can change their frequency of operation.

The unit is extremely small in size and, therefore, portable. The power supply is a 6-volt battery source made up of four AA penlight cells. Standard zinc-carbon cells give a battery life of approximately 10 hours if the instrument is left on continuously. Considerably longer battery life is obtained if alkaline cells are used. There is no provision for an ac power supply.

It is important to note that the condition of the batteries greatly affects the operation of the counter. This is particularly true of the upper frequency-measuring capability of the counter. The specifications indicate an upper frequency range of 3 MHz; typically, however, much higher frequencies can be measured if fresh batteries are used. A maximum frequency of up to 6 MHz can be accommodated with good batteries.

### Circuit Operation

Fig. 8-2 shows a complete schematic diagram of the Conar Model 202 frequency counter. Note that because the unit uses large-scale integrated circuits, the diagram is extremely simple. Most of the

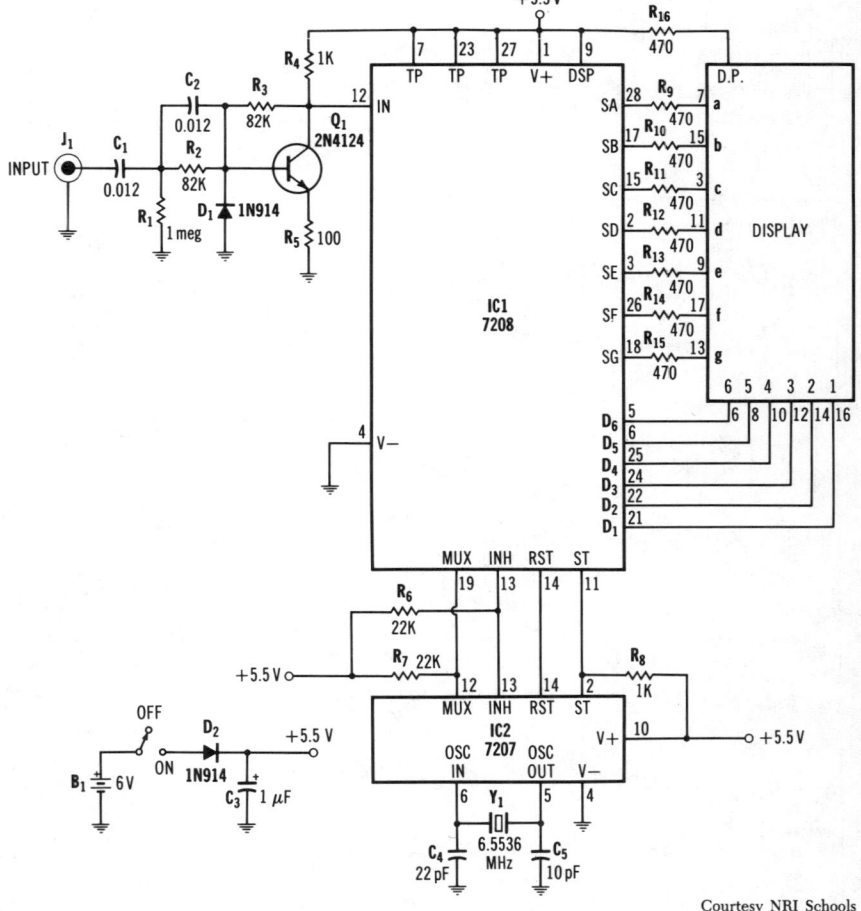

Fig. 8-2. Schematic diagram of Conar Model 202 frequency counter.

circuitry is contained within the Intersil 7208 integrated circuit labeled IC1. A detailed internal circuit diagram of this integrated circuit is shown in Fig. 8-3. The 7208 is a fully integrated decade CMOS counter housed in a 28-pin package.

Referring to Fig. 8-3, you can see that the 7208 LSI CMOS counter consists of seven cascaded bcd decade counters (labeled "÷10"). The input is first applied to the main gate and conditioned by a wave-shaping circuit before it is applied to the least significant counter on the left. The counter input is applied to pin 12 of the LSI counter chip. The input signal passes through the input resistance and a CMOS serial main gate.

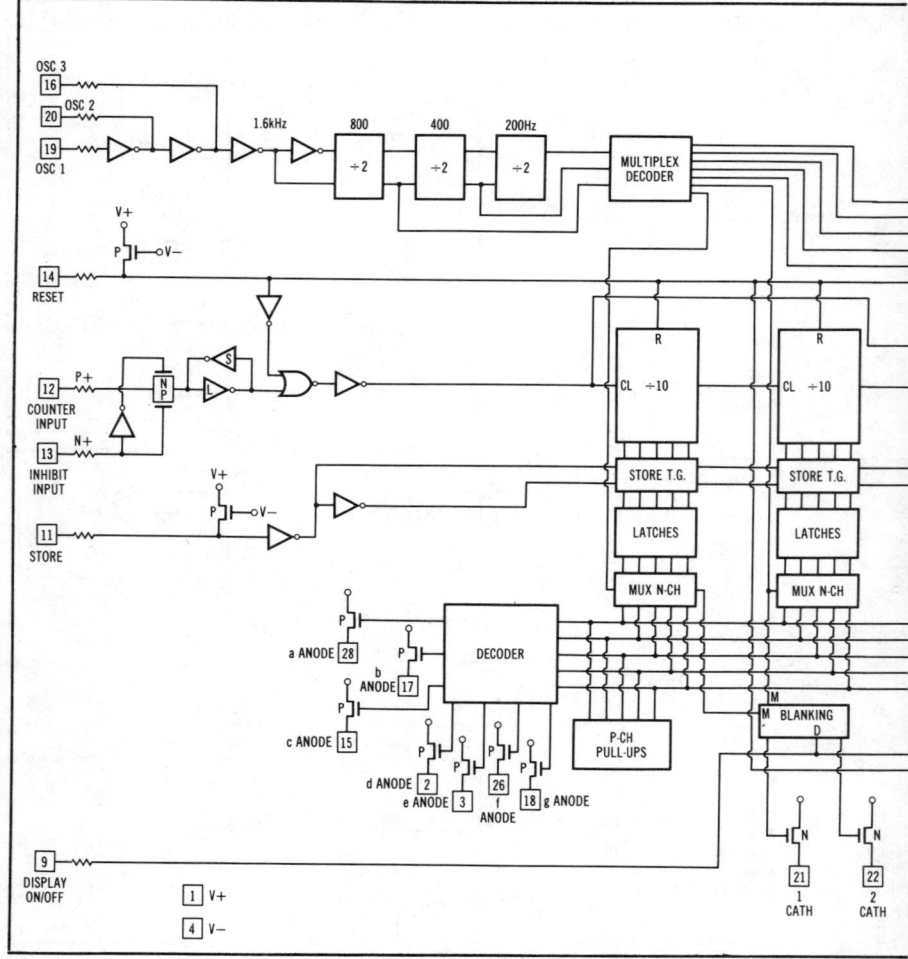

Fig. 8-3. Block diagram

A switch formed by p- and n-channel CMOS transistors opens or closes to either pass input pulses or inhibit them. This gate is controlled by the inhibit input on pin 13.

The output of the main gate is fed to a pair of back-to-back inverters labeled "S" and "L." These form a type of Schmitt trigger that is used to shape the input into a suitable logic signal. It is the hysteresis of this shaping circuit that effectively sets the input sensitivity. The input signal passed by the main gate and shaped by the trigger circuit is fed through a NOR gate and an inverter to the input of the first decade counter.

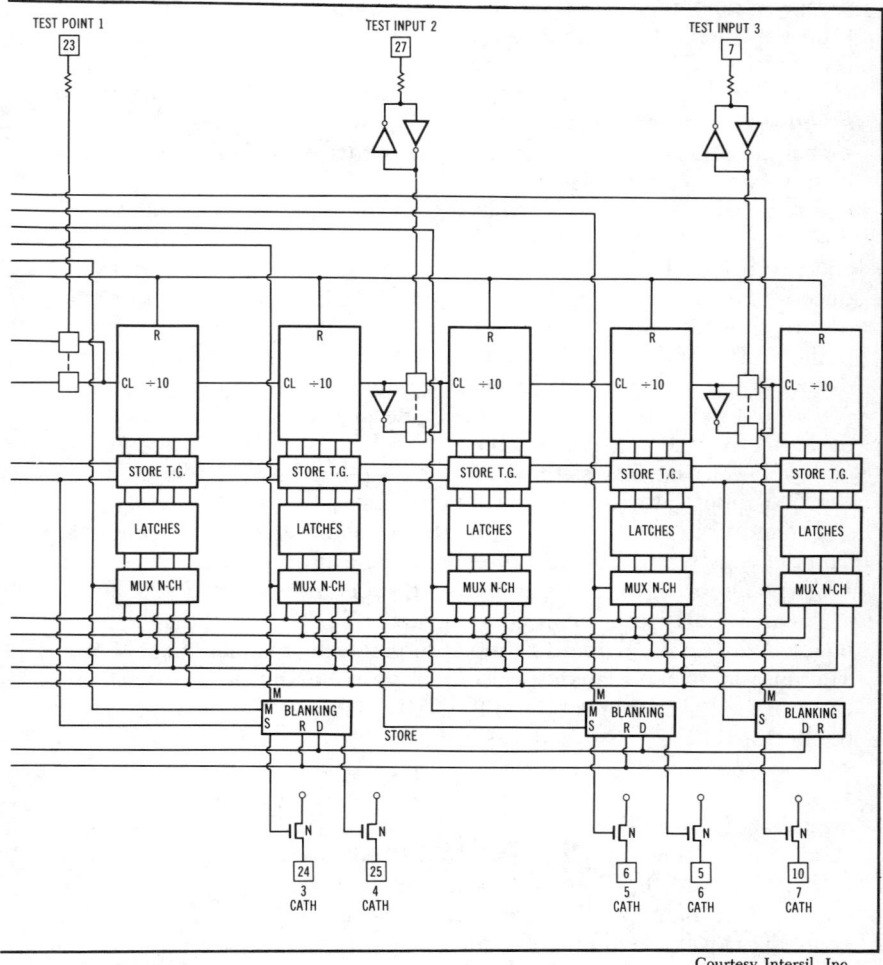

Courtesy Intersil, Inc.

of 7208 CMOS LSI counter.

Note that the bcd outputs of each decade counter are applied to a block labeled "Store TG." Store TG refers to a set of transfer gates that can be enabled or disabled. It is these gates that are enabled to allow the count in the decade counters to be transferred to a set of storage latches. The store input on pin 11 enables the gates and causes the bcd numbers in each counter to be transferred to the latches. Each box labeled "Latches" contains four flip-flops that form a bcd storage register.

The boxes labeled "MUX N-CH" at the output of the latches, one for each decade counter, are a set of AND gates that cause the outputs of the latches to be transferred to the decoder circuit. Since a multiplexed display arrangement is used, the output of only one set of latches at a time is applied to the decoder circuit. The desired MUX N-CH gates are enabled to apply the latch outputs to the decoder.

The decoder takes a standard four-bit bcd input and converts it into a seven-segment display output. The outputs of the decoder drive p-channel MOS transistors that are used to turn on and off the segments of the external LED displays.

The multiplex-decoder block in Fig. 8-3 determines which of the bcd digits from the various latches is passed to the decoder. A three-digit counter driven by an external 1.6-kHz signal operates the multiplex decoder and scans the MUX N-CH gates. As the 1.6-kHz input signal occurs, the three-digit counter is incremented and applies binary signals to the multiplex decoder. Only one of the seven outputs of the multiplex decoder is enabled at a time. The multiplex decoder, therefore, sequentially enables the MUX N-CH gates one at a time. When one set of MUX N-CH gates is enabled, the bcd number stored in the corresponding latches is applied to the decoder. The decoder converts the bcd input into the desired seven-segment output.

The multiplex decoder also sequentially enables the blanking circuits. The blanking circuits control the turn-off and turn-on of the individual LED readouts. Since there are seven bcd counters in the 7208, seven digits of display can be accommodated. These seven displays are enabled and disabled one at a time in sequence by the blanking circuits. The outputs of the blanking circuits drive n-channel MOS transistors that are used to enable or inhibit the LED readouts. These are shown at the bottom of Fig. 8-3. Keep in mind that only one of the LED displays will be enabled at a time. When an LED display is enabled, the appropriate bcd number is decoded, and the correct seven segments are illuminated. The Conar Model 202 has six digits of display. The most significant (seventh) digit in the 7208 is not used.

Because of this LED enabling and decoding process, the displays are not on continuously. However, since the displays are scanned and enabled at a very high rate of speed, to the user they *appear* as though they are on continuously. This is the principle of multiplexed display operation. The multiplexing system greatly reduces the amount of circuitry, the cost, and the power dissipation associated with displaying a large number of digits.

Referring back to the diagram of the Model 202 in Fig. 8-2, note the circuitry used to condition the input signal before it is applied to pin 12 of the counter IC. The input circuit uses ac coupling, which consists of C1 and R1. A single transistor amplifier, Q1, is used to provide isolation between the input signal and the counter IC. A small amount of gain (less than 10) is provided. Diode D1 is used along with R2 to clip very high levels of input signal to prevent damage due to overloads. The collector output of amplifier Q1 is applied to the 7208 counter input at pin 12.

The 7207 integrated circuit (IC2) contains the crystal oscillator, time-base divider, and control circuits for the counter. A 6.5536-MHz crystal is connected to the device. The internal clock generates a 6.5536-MHz signal which is, in turn, divided down by internal flip-flops. The result is applied to a series of logic circuits that generate the various control and gating signals for the counter. This unit also supplies the 1.6-kHz signal used to drive the display multiplex circuitry.

Fig. 8-4 shows the waveforms generated by the 7207 time-base circuit. These are the main-gate enabling signal, the reset pulse, and the store, or transfer, signal. The gate signal is applied to the inhibit input of the counter on pin 13. This is the signal that turns the main gate off and on. A 0.1-second gate interval is used in the Model 202.

Following the counter time-base input, a store, or transfer, signal pulse is generated. This pulse is applied to pin 11 of the 7208. This causes the count stored in the internal bcd counters to be transferred to the latches. Next, the 7207 time-base circuit generates a counter-reset pulse. This reset pulse is applied to pin 14 of the 7208 counter. This resets the bcd counters to zero and prepares them for another enable input from the 0.1-second gate. The cycle continues so that the frequency is repeatedly measured, transferred, and displayed.

Finally, note the simple battery power-supply circuit shown in Fig. 8-2. Four type-AA 1.5-volt penlight cells are connected in series to provide a 6-volt input. An off/on switch connects the battery to the supply circuit consisting of D2 and C3. Diode D2 simply provides reverse-polarity protection for the circuitry. When the cells are installed properly, the diode conducts and provides a voltage drop of approximately 0.5 volt. If the battery should be reversed, diode D2 will not conduct and, therefore, will not supply voltage to the circuitry. This prevents damage to the integrated circuits caused by accidental polarity reversal due to improper battery installation. Capacitor C3 provides a low-impedance ac path for the battery supply. The 5.5-volt output operates the integrated circuits, input amplifier, and LED displays.

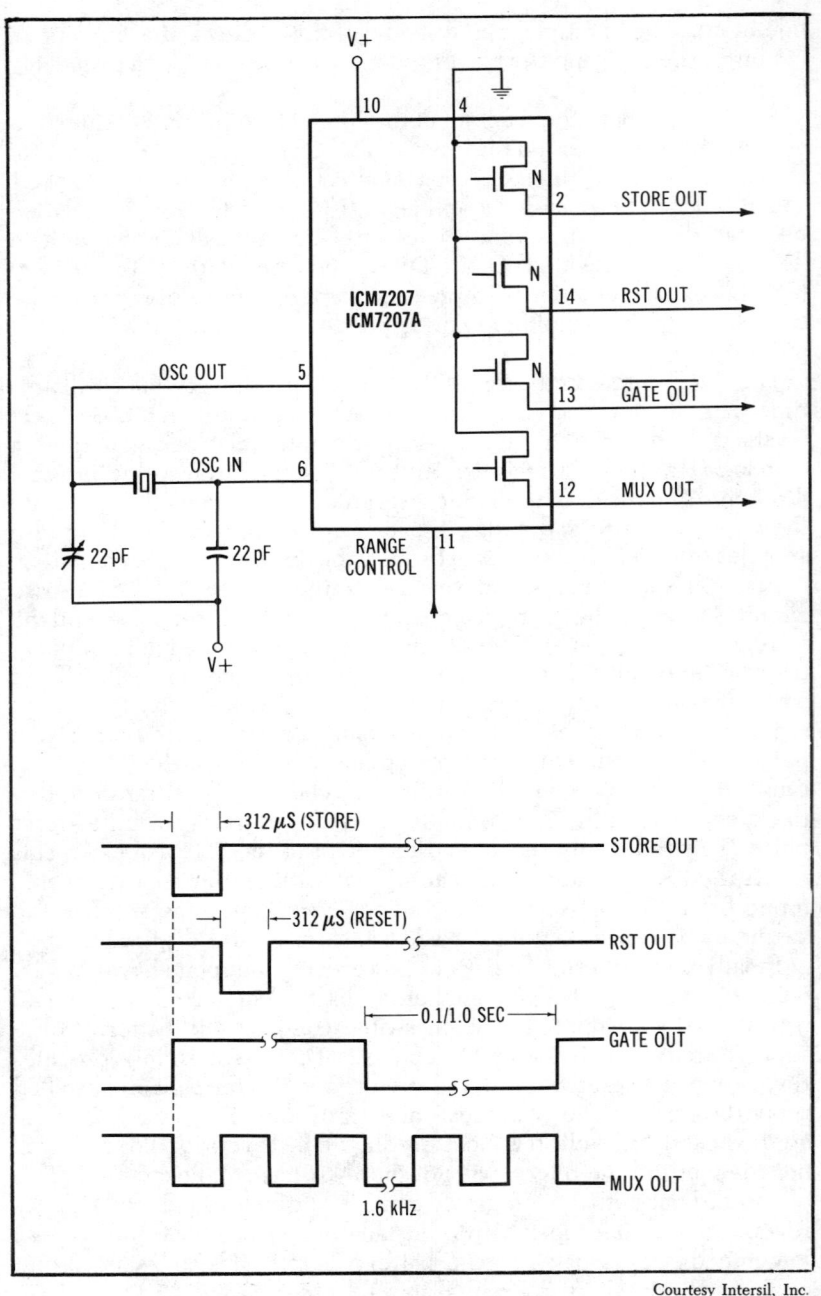

Fig. 8-4. Diagram and waveforms of 7207 clock oscillator/timing controller.

## LSI COMMUNICATIONS COUNTER

The Heathkit Model IM-2400 (Fig. 8-5) is a handheld, portable, battery-operated frequency counter used for rf measurements up to 512 MHz. The main application of this counter is in servicing and adjustment of communications equipment. The unit is useful for checking mobile radio transmitters and handheld transceivers in both commercial and amateur radio applications. The unit is supplied as a kit. As in the Model 202 discussed previously, the heart of this counter is an LSI chip.

Fig. 8-5. Heathkit Model IM-2400 LSI frequency counter.

Courtesy Heath Co.

## Chart 8-2. Specifications of Heathkit Model IM-2400

| | |
|---|---|
| Frequency Ranges: | 50 Hz to 50 MHz |
| | 40 to 512 MHz |
| Number of Display Digits: | 7 |
| Resolution: | 50-MHz range—0.1-sec time base, 100 Hz |
| | — 1-sec time base, 10 Hz |
| | 512-MHz range—0.1-sec time base, 1 kHz |
| | — 1-sec time base, 100 Hz |
| Time Base: | Frequency—10 MHz |
| | Temperature Stability—±10 ppm, 0°C to 40°C |
| | Setability—±1 ppm |
| Input Impedance: | 50 Hz to 50 MHz—1 megohm shunted by 24 pF |
| | 40 to 512 MHz—50 ohms |
| Sensitivity: | 50 Hz to 50 MHz—25 mV rms max, 10 mV rms typical |
| | 40 to 512 MHz—25 mV rms |
| Maximum Input: | 50 Hz to 50 MHz—150 V rms at 100 kHz to |
| | 10 V rms at 50 MHz |
| | 40 to 512 MHz—5 V rms |
| Power: | 6.25 volts (five AA nickel-cadmium rechargeable batteries) |

## Specifications

The specifications for the Model IM-2400 are given in Chart 8-2. The frequency coverage of this counter is 50 Hz to 512 MHz in two ranges, 50 Hz to 50 MHz and 40 MHz to 512 MHz. The counter has a seven-digit display.

The resolution of this counter depends on the time-base selection and the frequency range. Two time-base intervals, 0.1 second and 1 second, are selectable with a front-panel switch. With a 0.1-second time-base, the resolution is 100 Hz on the 50-MHz range and 1 kHz on the 512-MHz range. Selecting the 1-second time-base interval gives a resolution of 10 Hz on the 50-MHz range and 100 Hz on the 512-MHz range.

The heart of the time base is a 10-MHz crystal oscillator, which is actually part of the LSI counter chip. This circuit provides a temperature stability of ±10 parts per million (ppm) over the range of 0°C to 40°C. With an external trimmer capacitor, the frequency setability is ±1 part per million.

The IM-2400 counter has two separate input circuits, one for the 50-MHz range and the other for the 512-MHz range. The input impedance of the 50-MHz input is 1 megohm shunted by approximately 24 pF. Additional shunting capacitance is added by the cable. For the 512-MHz input, the impedance is a standard 50 ohms.

The sensitivity on the 50-MHz input is typically 10 millivolts rms. This can be as high as 25 millivolts on the average unit. Sensitivity on the 512-MHz range is 25 millivolts.

The unit is powered by five AA-size nickel-cadmium rechargeable batteries. Under load, these batteries generate 6.25 volts, which is used to run all of the internal circuitry. Under continuous operation, the batteries will last from 1 to 1½ hours. The batteries can then be recharged for a 14- to 24-hour period with the external ac-operated charger unit supplied with the counter.

The very small size and battery-operated nature of this counter make it fully portable and, therefore, handy for making frequency measurements at remote transmitter sites or on vehicles in which mobile radios are installed. An accessory whip antenna is available for the counter to permit rf-coupled frequency measurements as well as standard cable-connected measurements.

## Circuit Operation

The heart of the IM-2400 counter is an Intersil 7216D LSI counter circuit. This device is even more sophisticated than the 7208 LSI chip described earlier. Basically, it contains all of the functions of the 7208 and the 7207 time-base circuit combined plus other features.

The 7216D circuit consists of a high-frequency clock oscillator, a decade time-base divider, and an 8-decade data counter with storage latches. A seven-segment decoder, digit multiplexers, and 8-segment/8-digit driver circuits are used to operate external LED displays directly. The maximum frequency of operation is 10 MHz. The 7216D counter chip contains all of the circuitry used to perform frequency-measurement functions. With a 10-MHz external time-base crystal, gate times of 10 seconds, 1 second, 100 milliseconds, and 10 milliseconds are selectable.

A general block diagram of the 7216D LSI circuit is shown in Fig. 8-6. The main counter (labeled "$\div 10^8$") shown in the center of the diagram consists of eight cascaded bcd counters. The four lines from each of these bcd counters are fed to data latches. All of the related output-multiplexing gate circuitry is also contained in the data latches block. The four-line multiplexed output is applied to the decoder logic, which translates the bcd digit into the appropriate 7-segment output. The 7-segment outputs feed the segment drivers, which are transistors capable of directly operating the segments on external LED displays. The decimal point (D.P.) logic block determines the position of the decimal point and lights the appropriate decimal point in one of the eight LED readouts.

The 7216D counter chip also includes all of the time-base circuitry. The clock oscillator is shown in the upper left-hand corner of Fig. 8-6. An external 10-MHz crystal is connected to the chip along with several capacitors for adjusting the circuit. Alternatively, a 1-MHz crystal may also be used.

The output of the clock oscillator drives the time-base divider circuit

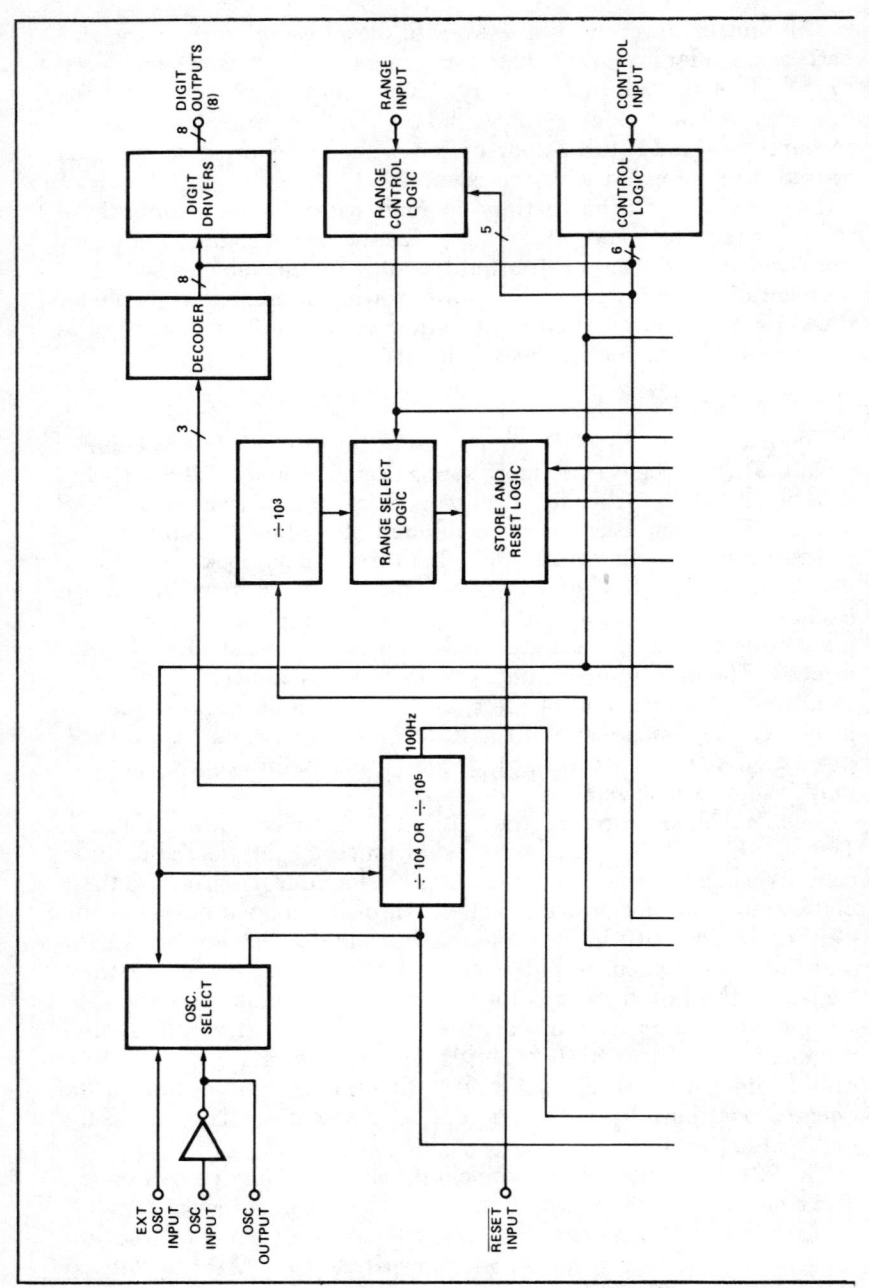

Fig. 8-6. Block diagram

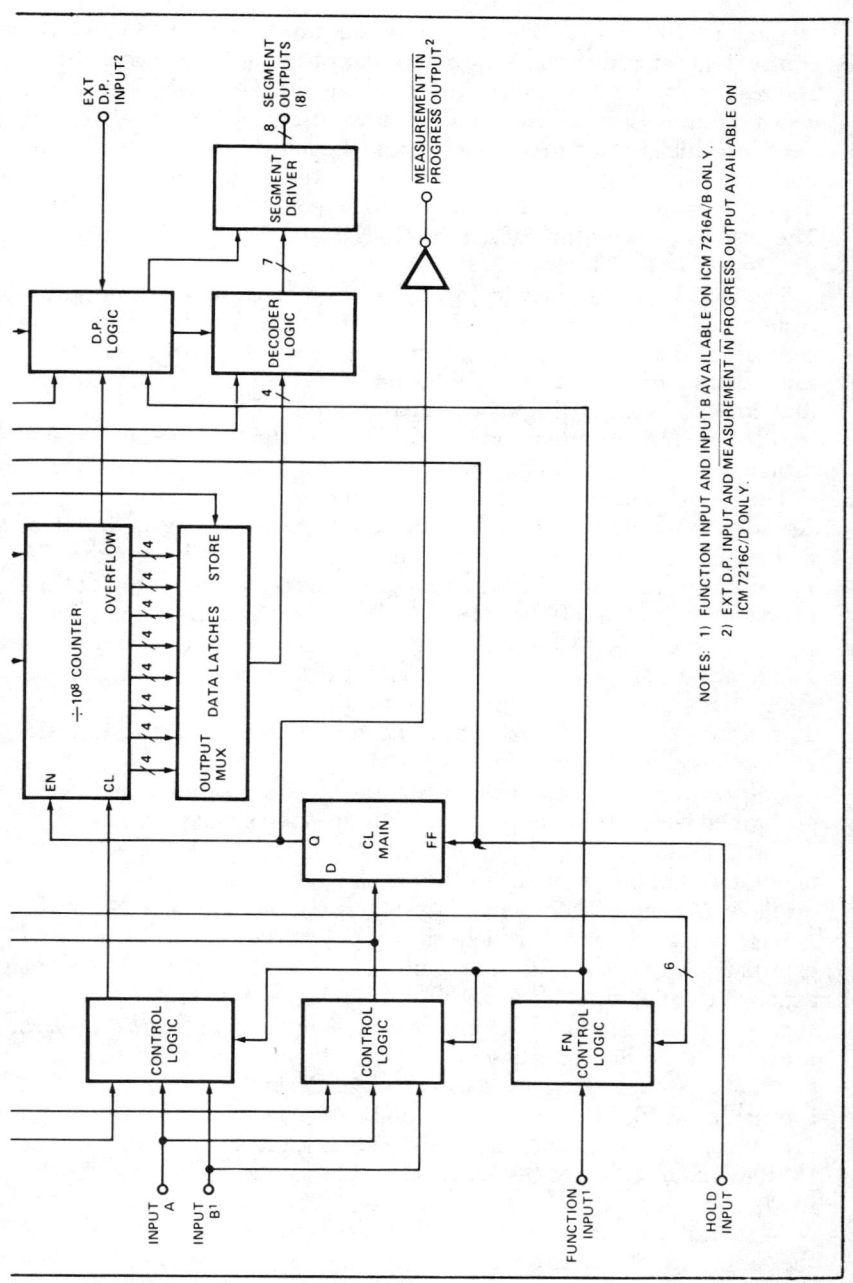

of 7216 CMOS LSI counter.

Courtesy Intersil, Inc.

labeled "÷ 10⁴ or 10⁵." The time-base outputs are used to drive the control logic circuits for generating the various time-base enable, store, and reset pulses for the main counter and latches. The time-base divider also generates three clock signals that are used to drive a 3-to-8–line decoder which is used to scan and enable the multiplexed external LED displays. The eight output lines from the decoder drive eight digit-driver transistors, which sequentially enable the external LEDs. The displays are multiplexed at a 500-Hz rate. With eight digits, the duty cycle is 12.5% for each digit.

The remaining circuitry in Fig. 8-6 is used to set up and control the counter. These circuits are used to determine the function, range, and control of the counter. Some versions of this counter, 7216A and 7216B, are capable of period, frequency ratio, time-interval, and totalize functions as well as frequency measurement. On the 7216D, only frequency measurement is available, so the "Function Input" designated in Fig. 8-6 does not exist on the 7216D.

The range-control logic is used to set the main-gate enable interval. As indicated earlier, the time base may be set to generate gate durations of 10 seconds, 1 second, 100 milliseconds, and 10 milliseconds. The range and control functions are time-multiplexed to select the desired characteristic. The time-base selection process is actually accomplished by connecting one of the eight digit output lines back around to the range input. The digit-driver outputs are labeled D0 through D7. Connecting D0, D1, D2, or D3 to the range input causes the 10-millisecond, 100-millisecond, 1-second, or 10-second gate interval, respectively, to be selected.

The control logic causes a number of control functions to be performed within the counter. One of these is the display test, in which all LED segments and decimal points are illuminated to verify their operation. Another function is the display-off mode, which blanks all displays and inhibits the counting operation. Another control is the 1-MHz–select function, which allows a 1-MHz crystal to be used externally with the same digit multiplex rate. The decimal point is adjusted to compensate for the use of the 1-MHz crystal rather than the standard 10-MHz crystal. An external-oscillator enable function is also provided. This allows the on-chip clock oscillator to be inhibited and an external time-base signal used. An external-decimal-point enable mode is provided so that a decimal point will be displayed whenever the digit driver connected to the external decimal point is activated. In the IM-2400 counter, only the external-decimal-point enable function is used.

As indicated earlier, the control logic blocks in Fig. 8-6 contain all of the gating circuitry used to generate the main-gate operation signals and the reset and store pulses for the main counter and the data latches.

Fig. 8-7 is a complete schematic diagram of the IM-2400 counter. The circuitry is divided between two circuit boards, the main circuit board and the display circuit board. The 7216D LSI counter chip is mounted on the display circuit board along with seven LED displays. The 10-MHz crystal and the related components are also mounted on this board. The counter communicates with the main circuit board, which contains the input circuits, prescaling counters, and related circuitry.

This counter has two separate input circuits. The upper input circuit handles signals from 50 Hz to 50 MHz. Signals between these frequencies are applied to input jack J1. Resistor R1 sets the input impedance to 1 megohm. The input is capacitively coupled through C1. Resistor R2 and diodes D1 and D2 form a clipping circuit that limits the amplitude of the signal to prevent damage. Transistor Q1 is a source follower that drives emitter follower Q2. This circuit has a high input impedance and a low output impedance. It provides the drive for integrated circuit U1.

Integrated circuit U1 is an ECL (emitter-coupled logic) line receiver circuit. It is made up of three differential-amplifier circuits labeled "A," "B," and "C." The first two of these line receiver stages are connected as linear differential amplifiers and provide a voltage gain of approximately 24 dB. The third stage, labeled "C," is connected as a Schmitt trigger and is used to shape and square the input signal. The output of U1 section C is an ECL logic signal. This is converted into a standard TTL logic signal by the translator circuit consisting of Q3 and the related components. The TTL-compatible output of emitter follower Q4 is the conditioned and shaped input signal that is ready for counting; it is applied to the multiplexer integrated circuit, U4.

The input circuit for high-frequency signals up to 512 MHz is shown in the lower left-hand corner of Fig. 8-7. High-frequency signals are applied to input jack J3. The input impedance is 50 ohms, which is made up of 100-ohm resistor R27 and all of the related circuitry that shunts it. The input signal is capacitively coupled to a diode bridge circuit made up of hot-carrier or Schottky barrier diodes D5 through D8. This bridge circuit provides limiting or clipping of the input signal to an approximate level of 300 millivolts peak-to-peak.

The signal output from the bridge clipper is applied to integrated circuit U2. This is a wideband differential amplifier circuit that provides a voltage gain of 20 to 30 dB through the 500-MHz range. The output voltage is self-limiting to 1 volt peak-to-peak.

The output of amplifier U2 is applied to integrated circuit U3, which is an ECL decade counter used as a divide-by-10 prescaler. Transistor Q5 converts the ECL logic levels to TTL logic levels. These are buffered by emitter follower Q6 prior to being applied to multiplexer IC U4.

The input circuit to be used is selected by range switch SW2. The left-hand section of this switch is used to apply 5 volts to either the low-frequency input circuit or the high-frequency circuit. When one

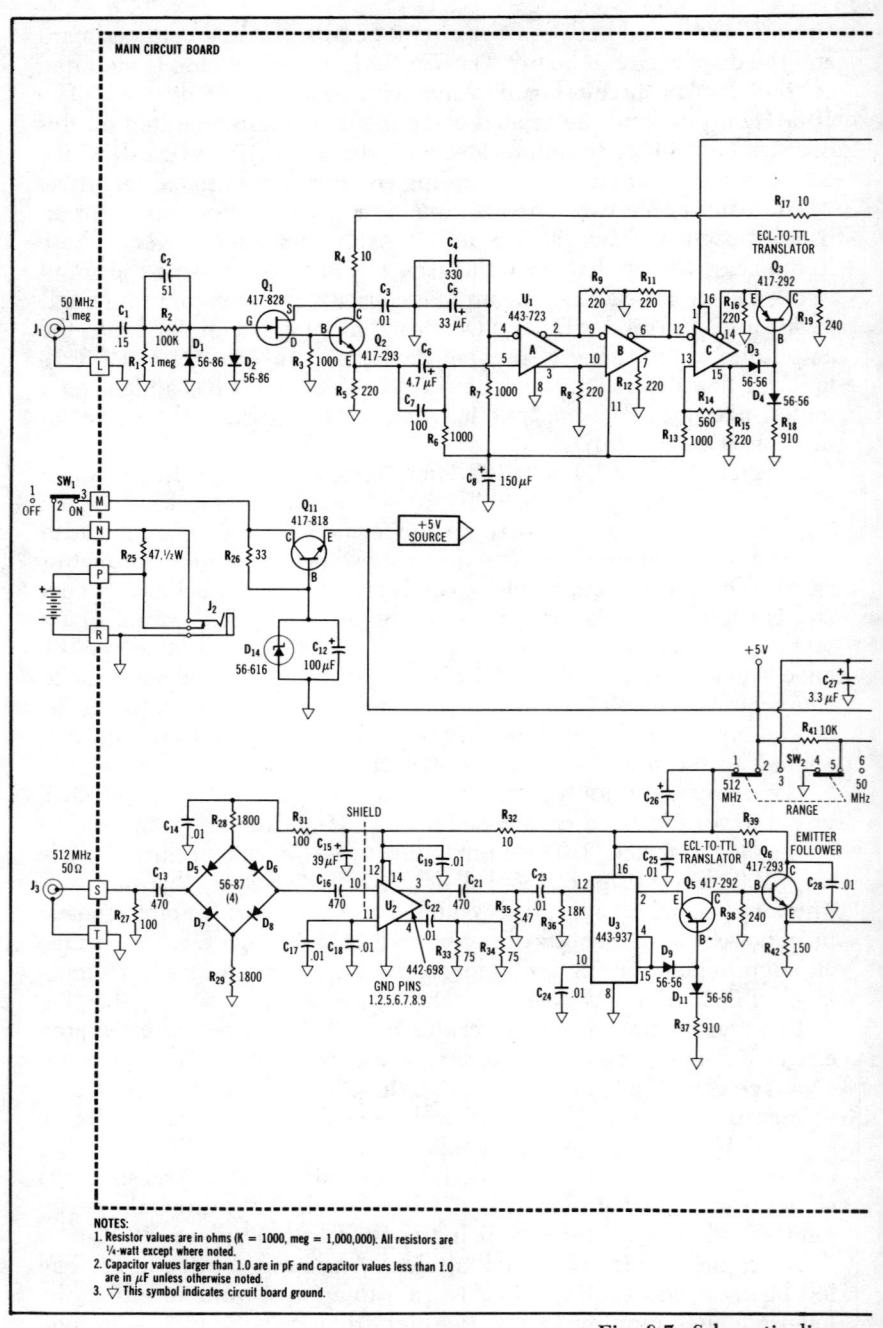

Fig. 8-7. Schematic diagram

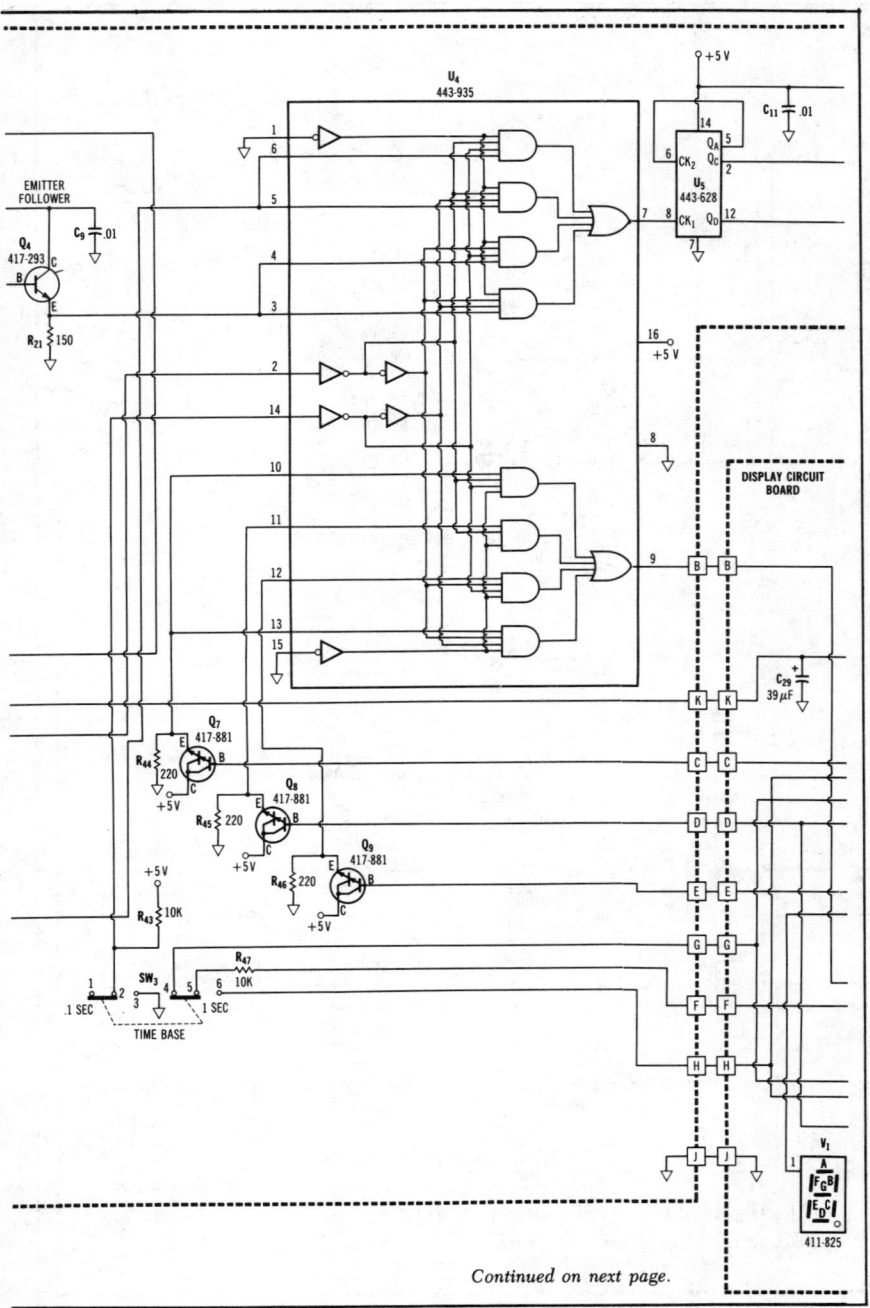

of the IM-2400 counter.

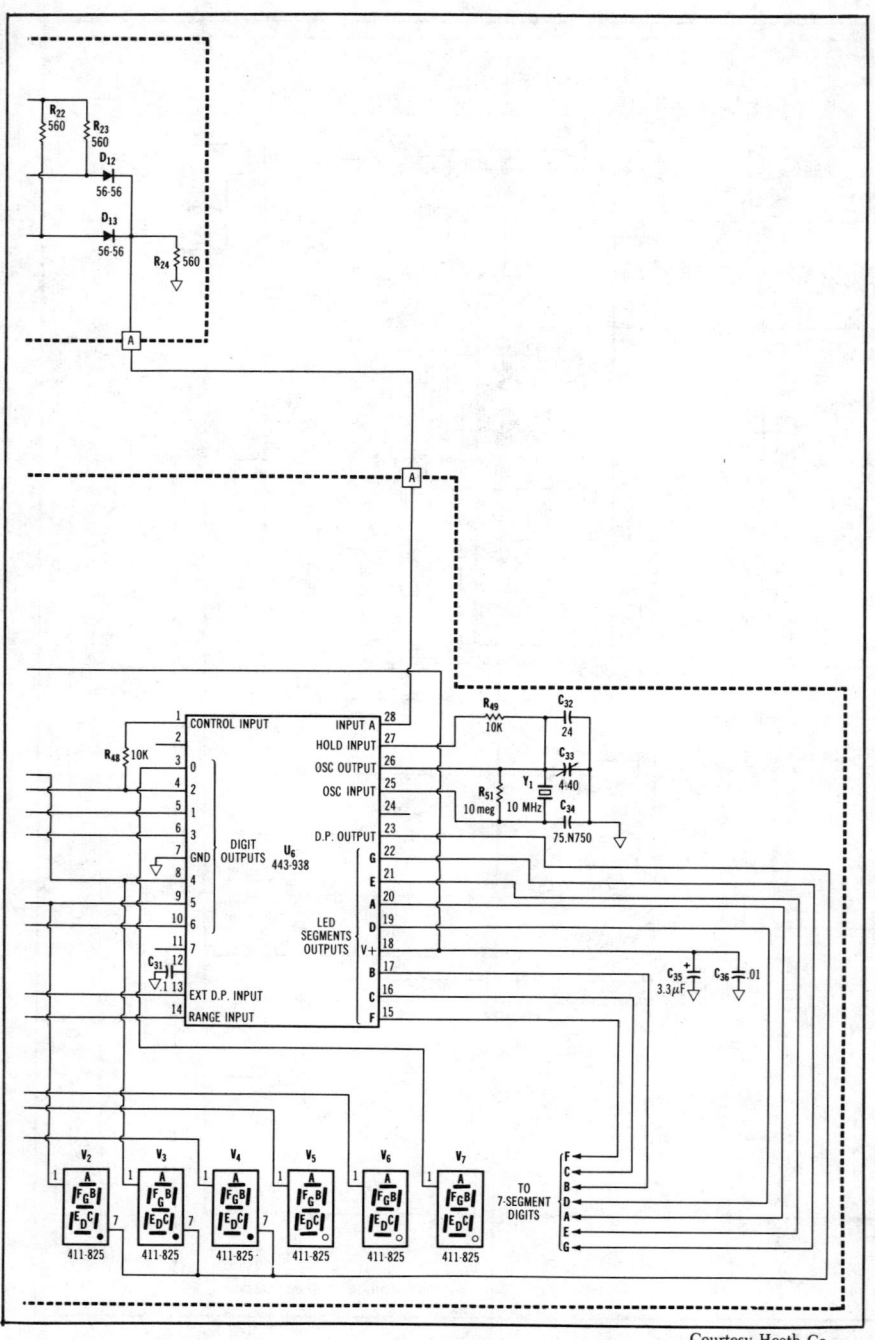

Fig. 8-7 (cont). Schematic diagram of the IM-2400 counter.

circuit is selected, the other is cut off completely. The right-hand section of SW2 applies a logic signal to pin 2 of multiplexer circuit U4. This logic signal selects whether the low frequency from emitter follower Q4 or the high frequency from emitter follower Q6 is passed through multiplexer U4.

Integrated circuit U4 is a dual four-channel digital multiplexer. The upper circuitry allows any one of the four input signals applied to pins 3, 4, 5, and 6 to be passed through to the output on pin 7. Which of these inputs is passed through to the output is dependent on the state of the logic signals at pins 2 and 14. The lower multiplexer is similar; one of the inputs from pins 10, 11, 12, and 13 will be passed through to the output at pin 9 depending on the states of the inputs at pins 2 and 14.

Note in Fig. 8-7 that the output from the low-frequency input circuits comes from emitter follower Q4. This signal is applied to pins 3 and 4 of multiplexer U4. The output from the high-frequency input circuit at Q6 is applied to multiplexer pins 5 and 6. The logic signal generated by the right-hand section of range switch SW2 is applied to pin 2 and, therefore, selects either the high- or low-frequency signal. One of these is passed through the upper section of the multiplexer and appears at pin 7. The output of the multiplexer is then applied to U5. This is a 74196 high-speed TTL bcd counter used as a decade prescaler. This TTL device is capable of counting at frequencies of up to 50 MHz. The normal output of the decade counter, $Q_D$, at pin 12 would ordinarily be applied directly to input A on pin 28 of the 7216D counter, U6. However, because of the waveshape limitations of the 7216D counter, an output signal with a longer duty cycle is required. The longer duty cycle is generated by an OR gate made up of diodes D12 and D13 and resistor R24. The $Q_C$ and $Q_D$ outputs of decade counter U5 are ORed together to produce a signal with a 60% duty cycle. This more than adequately meets the input requirements of the 7216D. This signal appears at pin 28 on U6.

The lower half of multiplexer U4 is used to select the position of the decimal point in the LED display. As indicated earlier, the 7216D counter provides for external decimal-point selection through pin 13. This is done by multiplexing the digit outputs and applying one of them to pin 13. The decimal point selected is that whose digit is enabled. For this display, LEDs V2, V3, and V4 contain the decimal point that must be properly illuminated depending on the settings of the time-base and range switches. One of these decimal points will be illuminated if the appropriate digit output, D3 at pin 6, D4 at pin 8, or D5 at pin 9, is connected back to the external-decimal-point input at pin 13. Which of the three digit outputs is selected is determined by the lower half of multiplexer U4. The D3, D4, and D5 digit outputs are applied to multiplexer U4 through emitter followers Q7 through Q9. These are Darlington pairs, which provide a very high input impedance for the

MOS signals from the 7216D counter. The emitter outputs are applied to the TTL inputs of the multiplexer. The output of the multiplexer at pin 9 is applied back to the external-decimal-point input at pin 13. Which of the three digit-select outputs is passed through the multiplexer is determined by the logic signals applied to pins 2 and 14 on U4. The right-hand section of range switch SW2 and the left-hand section of time-base-selection switch SW3 provide logic signals to make this selection.

The range input at pin 14 on the 7216D counter must also be connected to one of the digit outputs to select the appropriate time base. Both the 1-second and 100-millisecond gate periods are used in this counter. These are selected by connecting digit output 1 at pin 5 or digit output 2 at pin 4 on the 7216D back to the range input at pin 14. This is done through the right-hand section of time-base switch SW3.

As indicated earlier, the power supply for this counter consists of five built-in nickel-cadmium batteries. These generate a 6.25-volt supply, which is passed through a simple stabilizer circuit made up of Q11, D14, and the related components. Zener diode D14 along with the emitter-base junction of Q11 set the output voltage to +5 volts with a 6.25-volt input. The difference voltage is dropped across Q11. Five volts is applied to all of the remaining circuitry in the counter. Jack J2 is used to connect an external charger to the batteries. Plugging the charger into J2 causes the batteries to be charged while the external charging circuit operates the counter. Resistor R25 limits the charging current to the batteries.

## GENERAL-PURPOSE BENCH COUNTER

The Heathkit Model IM-2410 is an example of a standard general-purpose bench counter used for frequency measurements. This unit is typical of the many low-cost bench counters widely used in engineering labs, in service shops, and on hobbyists' and experimenters' benches. It is implemented with standard MSI TTL devices. No special LSI circuitry is used. The counter can be used for measuring frequencies as low as the audio range or radio frequencies up to 225 MHz, which includes the amateur 6-, 2-, and 1¼-meter bands. The counter is shown in Fig. 8-8.

### Specifications

The general specifications for this counter are given in Chart 8-3. It features a frequency range of 10 Hz to 225 MHz. An eight-digit LED display is used for frequency readout. Because of the additional display digit, this counter provides one order of magnitude better resolution than the seven-digit counter described in the previous section with the same time bases.

Courtesy Heath Co.
Fig. 8-8. Heathkit Model IM-2410 general-purpose bench counter.

### Chart 8-3. Specifications of Heathkit IM-2410

| | |
|---|---|
| **Frequency Range:** | 10 Hz to 225 MHz |
| **Number of Display Digits:** | 8 |
| **Resolution:** | 10 Hz to 50 MHz—0.1 sec time base, 10 Hz |
| | — 1-sec time base, 1 Hz |
| | 20 to 225 MHz —0.1-sec time base, 100 Hz |
| | — 1-sec time base, 10 Hz |
| **Time Base:** | Frequency—3.58 MHz |
| | Temperature Stability—±10 ppm, 0°C to 40°C |
| | Setability—±1 ppm |
| **Input Impedance:** | 1 megohm shunted by 24 pF |
| **Input Sensitivity:** | 25 mV rms |
| **Maximum Input:** | 150 V ac up to 100 kHz |
| | 5 V ac max from 160 to 225 MHz |
| **Power:** | 115/220 V ac, 50/60 Hz, 25 watts |

The time base in this counter is provided by a 3.58-MHz tv crystal. It has a temperature stability of ±10 parts per million over the range 0°C to 40°C. The frequency is setable to within ±1 part per million.

This counter has a single input connector. The input impedance is 1 megohm shunted by a capacitance of 24 pF. Any cable capacitance must be added to this. The input sensitivity is 25 millivolts rms over the frequency range.

The counter is powered by a standard 115- or 220-volt, 50- or 60-Hz power line. The unit is available only in kit form.

### Circuit Operation

Fig. 8-9 shows a complete logic diagram of the IM-2410 counter. The unit is made up primarily of standard MSI TTL devices. With the

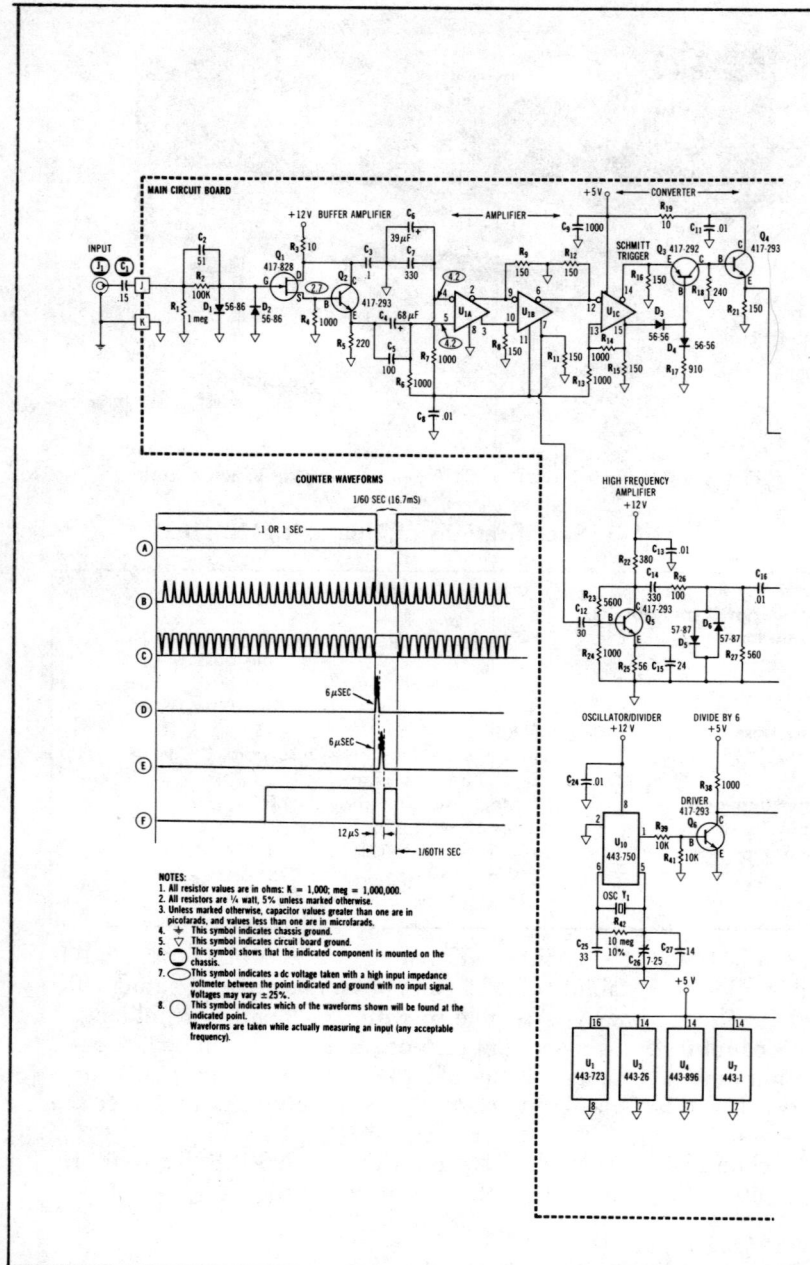

Fig. 8-9. Schematic diagram

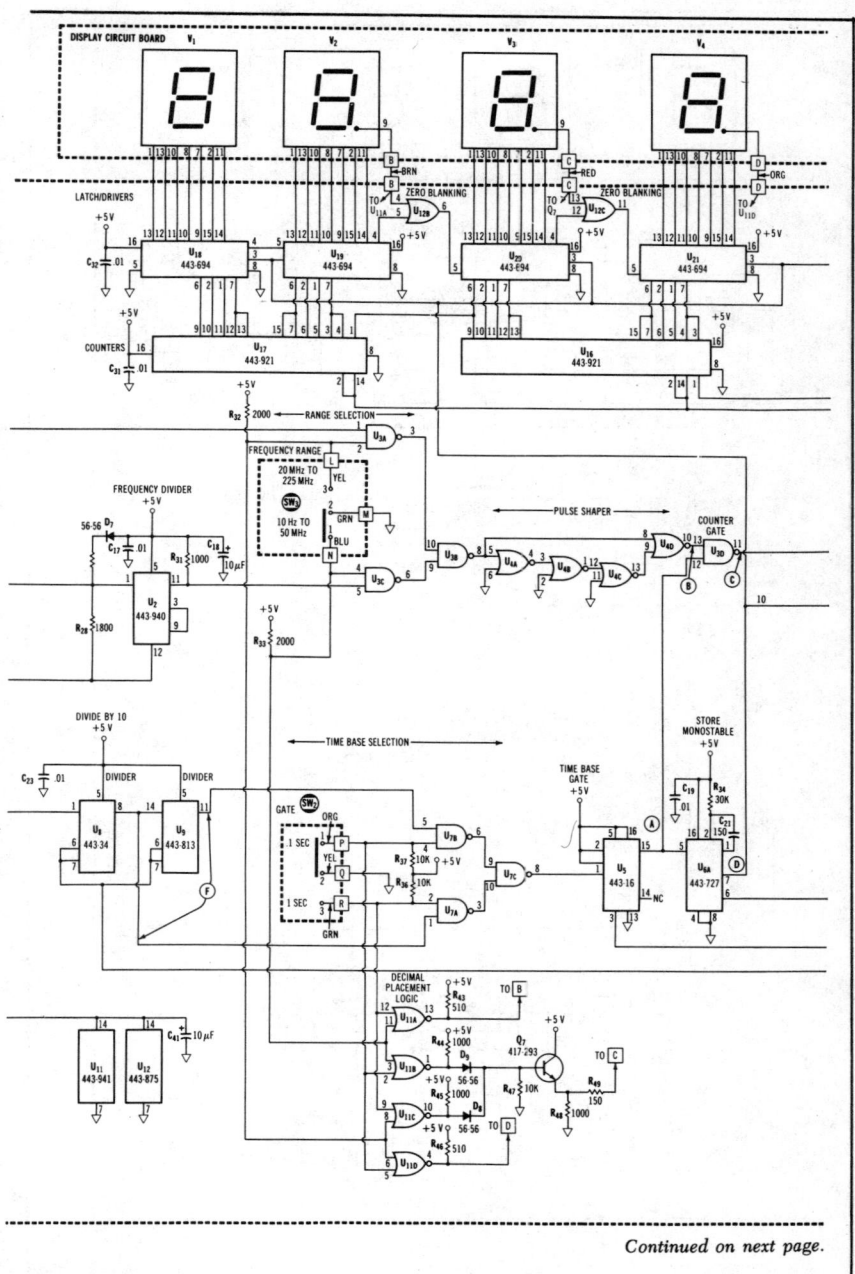

of the IM-2410 counter.

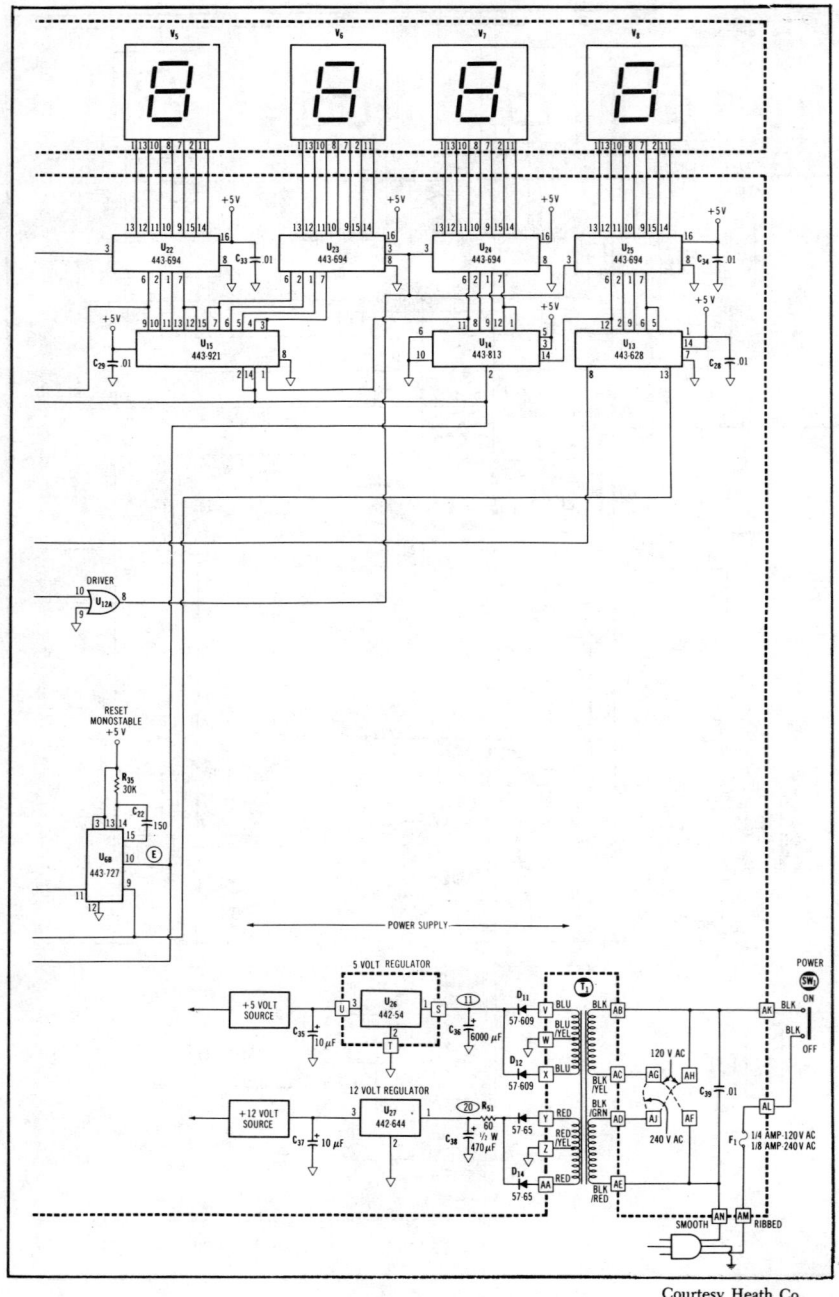

Fig. 8-9 (cont). Schematic diagram of the IM-2410 counter.

exception of the LED displays, all of the counter circuitry is mounted on a single printed circuit board.

The input circuit in the upper left-hand corner of the diagram is nearly identical to the input circuit in the IM-2400 counter previously discussed. Frequencies in the range 10 Hz to 50 MHz pass through this input-conditioning circuit and appear at the emitter of Q4. From there, the signal is further selected and conditioned prior to being counted.

For signals in the range 50 to 225 MHz, additional conditioning circuitry is used. Signals in that range still pass through the main input-amplifier circuits. However, the Schmitt trigger formed by integrated circuit U1C is not used. Instead, the high-frequency signal is tapped off at amplifier U1B and fed to an additional high-frequency amplifier, Q5. This amplifier provides an additional 20 dB of amplification. A clamping circuit made up of diodes D5 and D6 limits the output amplitude of the high-frequency amplifier. The signal is then coupled to integrated circuit U2, which is a high-frequency bcd counter used as a decade prescaler. At the maximum input frequency of 225 MHz, the output from U2 is 22.5 MHz.

The low-frequency output from Q4 and the high-frequency output from U2 are applied to the gates of integrated circuit U3. Gates U3A, U3B, and U3C form a two-input multiplexer circuit. Either the low-frequency or high-frequency output signal will be selected, depending on the position of the frequency-range switch, SW3. If the low-frequency range is selected, gate U3A will be enabled, and U3C will be disabled. If the high-frequency range is selected, U3C will be enabled, and U3A will be inhibited. The appropriate signal will then appear at the output of U3B.

The selected signal is then passed through a pulse-shaping circuit made up of gates in integrated circuit U4. Gates U4A, U4B, and U4C generate a 200-nanosecond delay which, when applied to gate U4D along with the selected input signal, generates a 200-nanosecond pulse that occurs on the negative-going transition of the input signal. This pulse is applied to the counter main gate, U3D. The 0.1- or 1-second time-base gating signal is also applied to main gate U3D.

The output of the main gate is applied to the least-significant-digit bcd counter U13. This is a 74196 TTL bcd counter capable of counting at a maximum rate of 50 MHz.

The output of the least-significant bcd counter is applied to the next bcd counter in sequence, U14. This is a 74LS90 bcd decade counter. Its output is applied to the remainder of the counter chain, which consists of U15, U16, and U17. Each of these devices is a 74LS390 dual decade counter.

The bcd outputs from the eight decade counters are applied to integrated circuits U18 through U25. These are 9368 TTL latch-decoder-drivers. Each circuit contains a four-bit storage register, a bcd–to–seven-segment decoder, and the related digit-driver circuitry. The outputs of these devices drive the LED displays.

Now let us take a look at the time-base and control-logic circuitry. The time base consists of a digital clock oscillator, U10. The oscillator is controlled by the 3.58-MHz crystal. Integrated circuit U10 also contains a frequency-divider circuit that generates a 60-Hz output. This is applied to transistor Q6, which drives integrated circuit U8, a 7492 TTL counter that divides by 6. Dividing the 60-Hz time-base signal by 6 produces a 10-Hz time-base signal. The output of U8 is the 10-Hz signal that creates the 100-millisecond gating interval. This signal is applied to gate U7A. The output of U8 is also applied to the input of integrated circuit U9, a 74LS90 bcd decade counter that is used to further divide the frequency by 10. This division creates a 1-Hz signal that produces the 1-second time-base gate. This signal is applied to gate U7B.

Gates U7A, U7B, and U7C form a two-input digital multiplexer. The time-base signal to be selected is determined by gate (time-base) switch SW2. When gate U7B is enabled, the 1-second time base is selected. When gate U7A is enabled, the 0.1-second time base is selected. The selected time-base signal is passed through gate U7C. It is then applied to integrated circuit U5, which is a JK flip-flop the output of which is the gating signal that is applied to the counter main gate, U3D.

The time-base signal from flip-flop U5 is also applied to integrated circuit U6, which is a dual one-shot multivibrator. The first section, U6A, generates a 6-microsecond pulse that is used to transfer the count in the bcd counters to the storage latches. Immediately following this pulse, the second one-shot multivibrator is triggered and generates a 6-microsecond reset pulse that clears all of the bcd counters, readying them for the next count sequence. Note that the output of the second one-shot multivibrator, U6B, in addition to generating the reset pulse is also fed back to the reset inputs on time-base dividers U8 and U9. This resets the time-base generation cycle and causes the counting cycle to repeat. The counting waveforms are illustrated in Fig. 8-9. The 0.1- or 1-second gate interval is followed by a store or transfer pulse and the reset pulse. The measurement cycle then begins again.

Integrated circuit U11 contains four gates that determine which decimal point in the display is illuminated. The decimal points in LEDs V2, and V3, and V4 are lighted according to the time-base selection and the frequency range. The inputs to the U11 gates come from the range and time-base selection switches. This logic is then used to turn on the decimal point appropriate to those selections.

Finally, the power supply is shown in the lower right-hand corner. It contains a dual-primary power transformer that may be wired for either 120-volt or 240-volt operation. The transformer also has two center-tapped secondaries. With the appropriate rectifier diodes, two full-wave rectifier circuits with capacitor filters are formed. The upper circuit generates approximately 11 volts across filter capacitor C36. This is applied to a 5-volt integrated-circuit regulator U26, which generates the 5 volts to drive most of the integrated circuits. The lower full-wave rectifier circuit generates approximately 20 volts across filter capacitor C38. This is applied to 12-volt voltage regulator U27. The 12-volt supply is used to operate the input circuitry and time-base oscillator U10.

## UNIVERSAL COUNTER/TIMER

The Fluke 7261A counter is an example of a universal instrument for measuring time and frequency. It uses MSI and LSI integrated circuits. In addition to its ability to measure frequency, it can perform virtually any of the time measurements discussed elsewhere in this book. It also features a number of interface options.

In this section, we will concentrate on the specifications and features of the counter rather than circuit operation. The circuits are so numerous and complex that it would take a book nearly this size to explain them completely. For that reason, we will emphasize only the general operation with block diagrams. Fig. 8-10 is a photograph of the Model 7261A counter/timer.

### Specifications and Features

The main feature of the Fluke 7261A counter is its ability to make a variety of time measurements in addition to measuring frequency. The specifications are summarized in Chart 8-4.

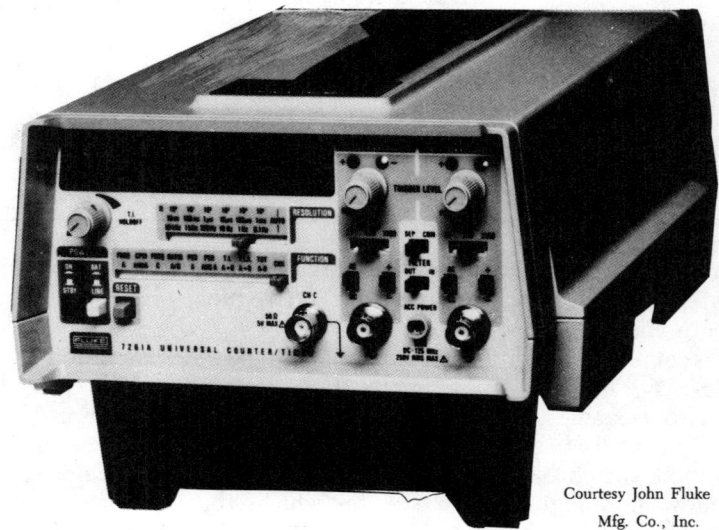

Courtesy John Fluke Mfg. Co., Inc.

Fig. 8-10. Fluke Model 7261A universal counter/timer.

## Chart 8-4. Specifications of Fluke 7261A

| | |
|---|---|
| Frequency Range: | 0 to 125 MHz (dc coupled, both channels) |
| Number of Display Digits: | 8 |
| Resolution: | 0.1 Hz to 10 kHz in decade steps |
| Time Base: | Frequency—10 MHz |
| | Temperature Stability—±5 ppm, 0°C to 50°C |
| | Aging Rate—±5 × $10^{-7}$/month |
| Input Impedance: | 1 megohm shunted by 55 pF (both channels) |
| Coupling: | ac or dc |
| Attenuator: | ×1, ×10, ×100 |
| Filter: | 100 kHz low-pass, switchable |
| Input Sensitivity: | 10 mV rms, 0 to 50 MHz |
| | 15 mV rms, 50 to 100 MHz |
| | 35 mV rms, 100 to 125 MHz |
| Trigger: | +1.5 V to −1.5 V |
| | Slope + or −, switchable |
| Maximum Input: | 250 V rms to 50 kHz |
| | 5 V rms 1 MHz to 125 MHz |
| | Linear operating range +2.5 V to −2.5 V |
| Modes: | Frequency |
| | Frequency (counts per minute) |
| | Frequency ratio A/B |
| | Period |
| | Period average |
| | Time interval |
| | Time interval average |
| | Totalize |
| | TI hold-off |
| Options and Accessories: | Third input channel front end |
| |    50 to 520 MHz, 10 mV rms, 50 ohms |
| | Data output interface, serial bcd of |
| |    all digits, TTL levels |
| | IEEE 488 interface |
| | Battery pack |
| | Higher-accuracy time bases |
| |    2 ppm |
| |    1 ppm |
| |    0.01 ppm |

This counter has two independent input channels, one referred to as channel A and the other as channel B. The frequency-measuring range for both channels is 0 to 125 MHz.

This counter has eight LED display digits. The time-base frequency is 10 MHz. For frequency measurements, the time base is fully selectable with a front-panel switch. As a result, the resolution ranges from 0.1 Hz to 10 kHz in decade steps, depending on the setting of the time-base, or resolution, switch. A separate position on the time-base selection switch provides autoranging, a mode in which the internal circuitry automatically selects the time-base range that optimizes both resolution and measurement time.

The input impedance for each of the two channels is 1 megohm shunted by 55 pF. Either ac or dc (direct) coupling may be selected with a front-panel switch. Each input channel also has a three-step attenuator built-in. A front-panel switch allows the input signal to be attenuated by a factor of 1, 10, or 100, depending on the magnitude of the input signal. The idea is to reduce the amplitude of high-voltage signals into the range best accommodated by the input amplifiers. The maximum linear operating range of the input circuits is +2.5 to −2.5 volts.

The input sensitivity for each channel varies with the frequency. The best sensitivity is 10 millivolts rms, which is obtained in the range 0 to 50 MHz. In the range 50 to 100 MHz, the sensitivity is 15 millivolts rms. Between 100 and 125 MHz, the input sensitivity is 35 millivolts rms.

Each input channel has an adjustable trigger-level control. This is a calibrated potentiometer that allows you to adjust the trigger level over a wide range. With the control in the maximum counterclockwise position, the trigger level is set to zero. This is the most sensitive trigger position. The trigger level can be adjusted anywhere in the range between +1.5 and −1.5 volts. A front-panel switch is used to select whether the trigger occurs on the positive-going or the negative-going slope.

Another input-channel feature is a built-in 100-kHz low-pass filter. In making measurements below 100 kHz, it is often desirable to switch in this filter and thus eliminate all high-frequency signals, which can sometimes cause noise and false triggering and thus measurement errors.

A BNC connector is provided for the input to each channel. A separate/common switch allows the two input channels to be used separately or to be connected together as required for some of the more sophisticated time-interval measurements.

The 7261A counter also makes provisions for measurements beyond the normal 125-MHz range of input channels A and B. An optional high-frequency input circuit is available at extra cost. This option simply plugs into the main counter chassis. A front-panel BNC connector is already provided. This input is designated channel C. With this option installed, the measurement capability is extended from 50 to 520 MHz. The input impedance is 50 ohms. This special high-frequency option contains a fused input and its own amplification and trigger circuitry. A decade prescaler is also contained within this option to bring the high input frequencies down within the range of the normal input channels.

The various measurement modes of the 7261A counter include frequency, frequency in counts per minute, frequency ratio, period, period average, time interval, time-interval average, totalize, and TI hold-off. Virtually any time or frequency characteristic can be measured with one of these modes.

This counter also features two optional interface circuits. These extra-cost items allow the counter to be connected to external equipment.

Finally, the 7261A counter, while primarily a bench instrument, also can be made portable. This is done by installing an optional battery pack.

## Counter Operation

Fig. 8-11 shows a general block diagram of the 7261A counter/timer. It is similar to the other counters discussed in this chapter, but in general it is much more sophisticated and complex because of its additional measuring capabilities. We will look at each of the blocks in the diagram in more detail.

Channels A and B are identical signal-conditioning input circuits. A more detailed diagram of this input circuit is shown in Fig. 8-12. It consists of an ac- or dc-coupled input with a 1-megohm impedance. A three-position attenuator switch permits high-amplitude signals to be reduced to a level more compatible with the signal-conditioning circuitry. A diode clamp circuit protects the input circuitry. A field-effect-transistor input buffer is used to isolate the input circuitry from the input amplifier. This amplifier is really a comparator circuit. The other input to the comparator is the voltage from the trigger-level control. The input amplifier is followed by a Schmitt trigger that shapes the input signal. The hysteresis of the Schmitt trigger can be adjusted by changing an internal resistor value. The output of the Schmitt trigger is applied to an ECL driver circuit that makes the Schmitt-trigger signal compatible with the remaining counting circuits. Note that LED indicators are used at the output of the ECL driver to indicate when positive or negative triggering takes place.

The channel C input block contains all of the circuitry associated with the optional high-frequency input. This input circuitry has a 50-ohm input impedance and its own signal conditioning, amplifier, trigger circuit, and prescaling decade counter.

The block labeled "Counter, Latch, & MUX" contains the main counting unit. A more detailed block diagram of this is shown in Fig. 8-13. The six most significant decade counters are contained within this chip. These are labeled "5-Digit BCD Counter" and "Overflow Counter." The outputs of these counters are applied to storage latches in the block labeled "7-Digit Latch" and then to a multiplexer circuit designated "7-Digit MUX." The scan clock and static scan counter enable the multiplexer so that the bcd digits from the storage latches are sequentially applied to the data output lines through the output buffers. The static scan counter also supplies strobe signals to the LED displays. The bcd data outputs and the strobe signals are fed to the display circuitry. The display circuitry contains the decoders that convert bcd coding to seven-segment coding; it also contains the LED drivers.

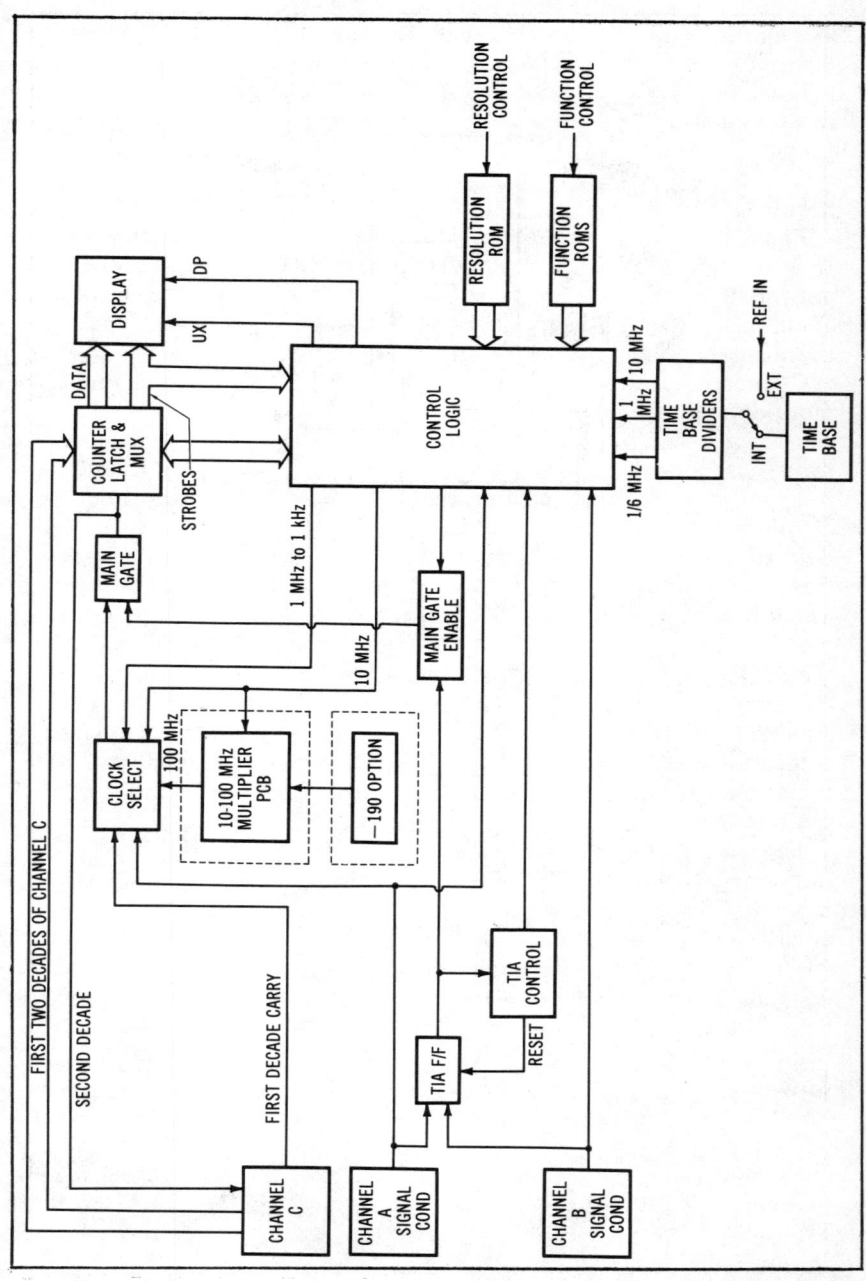

Courtesy John Fluke Mfg. Co., Inc.

Fig. 8-11. Functional block diagram of 7261A counter/timer.

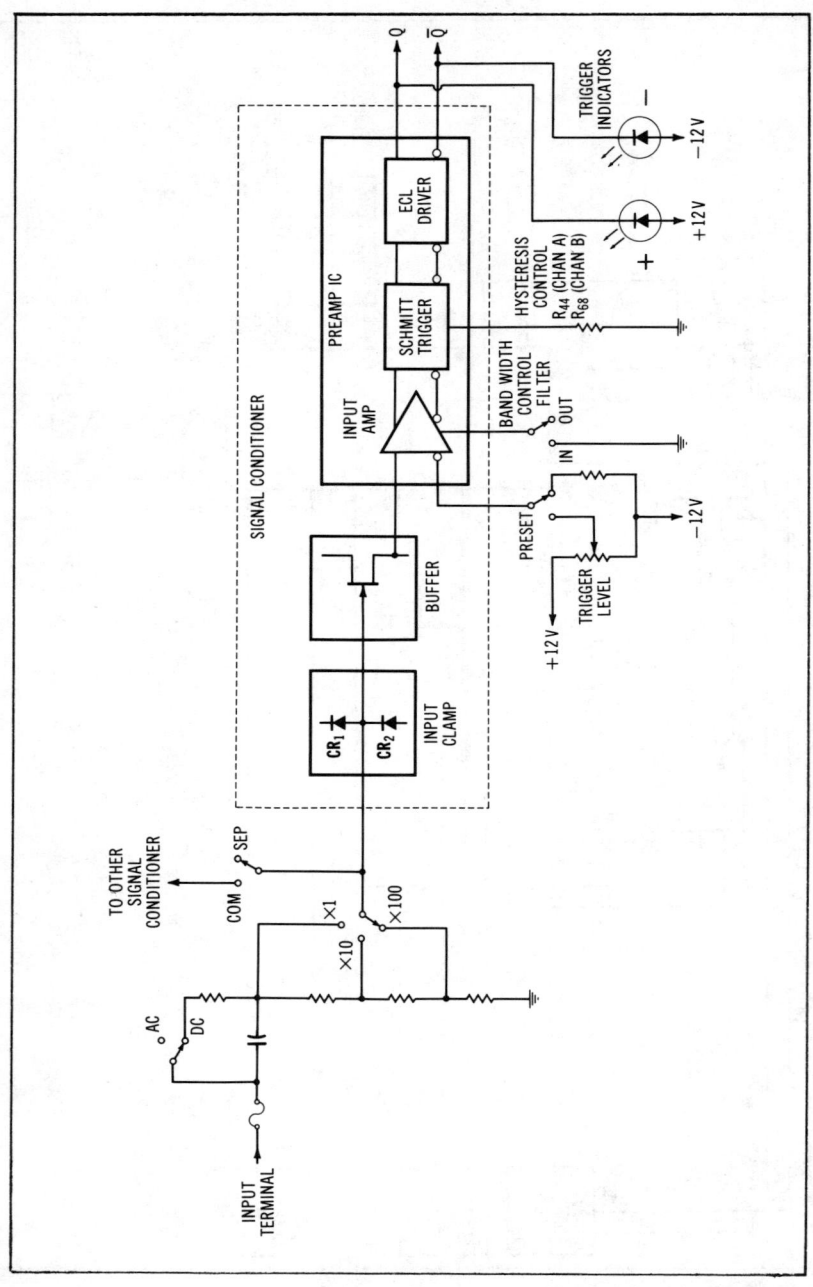

Courtesy John Fluke Mfg. Co., Inc.
Fig. 8-12. Input circuit of 7261A counter.

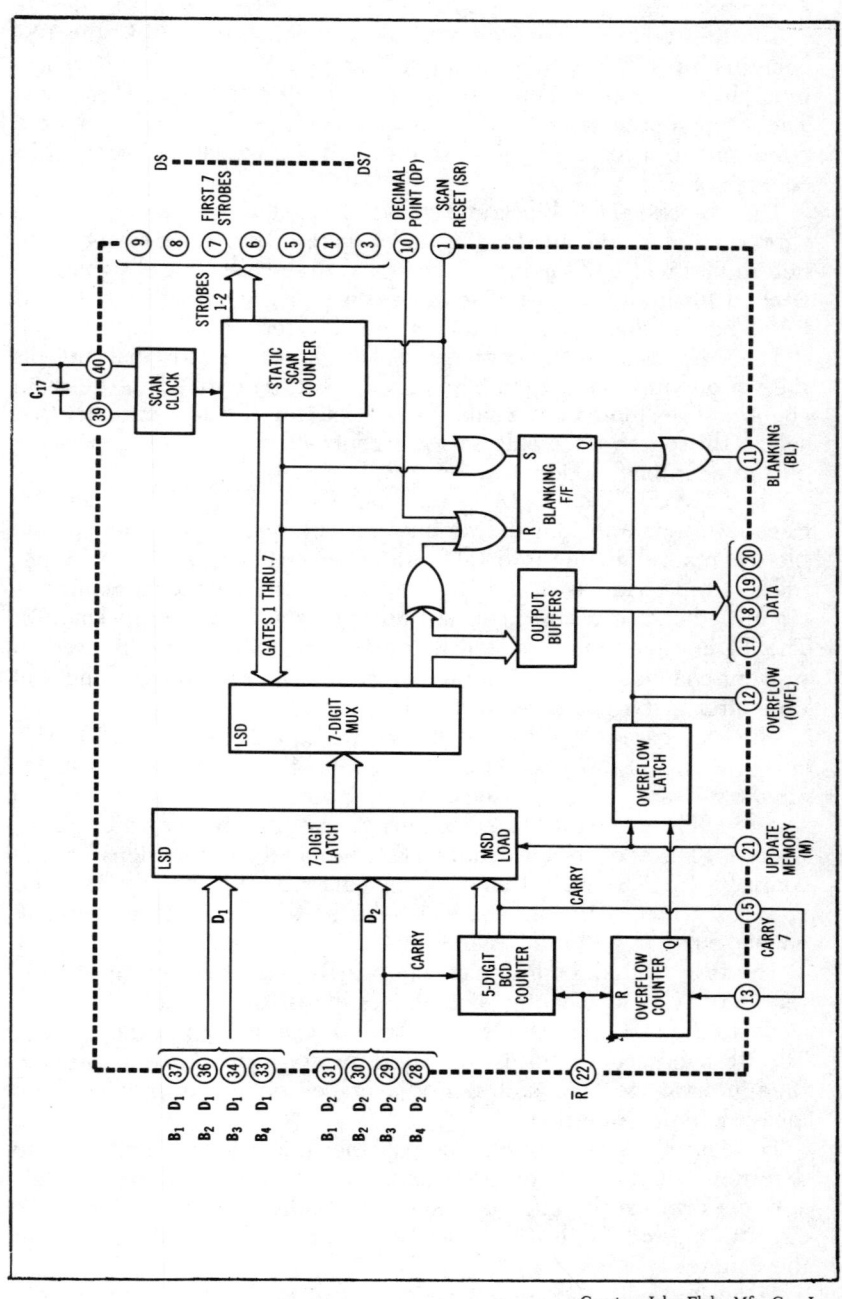

Courtesy John Fluke Mfg. Co., Inc.

Fig. 8-13. Counter, latches, and multiplexer.

The first two input bcd decades are not contained within the LSI counter chip in Fig. 8-13, but they are part of the counter circuitry. The input decade counter is made up of ECL and TTL high-speed flip-flops. The second decade is an MSI TTL decade counter. The outputs of these two input decades are applied to the digit storage latches in the LSI counter of Fig. 8-13.

The "Clock Select" block in Fig. 8-11 is a set of gates that route the various timing signals to the main gate. The clock-select logic determines which of the input signals and time-base control signals are applied to the main gate. These, of course, vary with the particular frequency or time measurement mode selected.

The "Main Gate" block contains all of the gating circuitry that controls the various input and time-base signals that are applied to the main counter. The "Main Gate Enable" block contains the logic circuitry that selects the source of the pulses used to control the main gate, depending on the counter function.

The "TIA F/F" and "TIA Control" blocks in Fig. 8-11 contain the circuitry used for making the various time-interval measurements. One block contains the time-interval flip-flop, which is set and reset by input pulses from channels A and B. The TIA control section contains the circuitry that controls the main gate during certain time-interval modes. The TI hold-off circuitry, which contains a timer circuit, is used to prevent channel B from generating a stop pulse for a preset length of time after a start pulse on channel A occurs.

The time base is the clock oscillator circuit that generates a 10-MHz signal. Fig. 8-14 shows a schematic diagram of the basic clock oscillator circuitry. Transistor Q2 is used in a Pierce oscillator, and Q1 is used in a source-follower circuit. The clock drives the time-base divider chain, which is a series of decade counters that supply the timing signals to the control logic. The main time-base outputs are 10 MHz, 1 MHz, and $1/6$-MHz. The $1/6$-MHz time-base signal is used to generate a main-gate enable pulse of 600 milliseconds.

The 10 − 100 MHz multiplier is a phase-locked-loop multiplier (see reference XIII in Chapter 4) that multiplies the 10-MHz time-base signal to 100 MHz. This 100-MHz signal has a period of 10 nanoseconds. The 10-nanosecond pulses are used in the period, time-interval, time-interval-average, and period-average modes to provide high measurement resolution.

The control logic block contains all of the various logic circuits used to determine the function or mode and the resolution. The control logic also performs such functions as decimal-point determination. This circuitry interacts with and otherwise controls all of the other blocks in the counter.

Associated with the control logic are two read-only memories, the resolution ROM and the function ROM. The front-panel function-con-

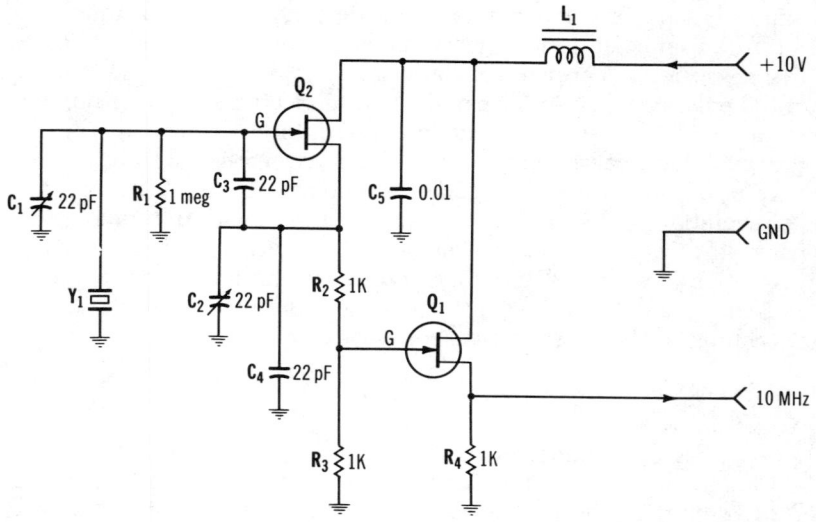

Fig. 6-14. Schematic diagram of 10-MHz time-base oscillator.

crol switch operates the function ROM, which in turn generates a group of digital signals that operate and interact with the control logic. The same is true of the resolution ROM. Depending on the front-panel time-base switch setting and other inputs, the resolution ROM generates logic signals that are used to enable or disable various circuits in the control logic. The ROMs are simply a compact and organized method of replacing what would otherwise take a significant amount of random combinational logic.

Now let us consider some of the specific operating modes of this counter. Frequency measurement is accomplished by applying the signal to be measured to channel A. The counter will accumulate pulses for one of several gate times, including 100 microseconds, 1 millisecond, 10 milliseconds, 100 milliseconds, 1 second, and 10 seconds. These gate times are selected by the front-panel resolution switch. If the auto position is selected, auto-ranging is used. In any case, the decimal point is automatically located so that the counter displays either kilohertz or megahertz.

Frequencies from dc to 125 MHz can be measured on channel A. For frequencies beyond this, the counter must have the channel C 520-MHz input option.

Another frequency-measurement mode is that designated cpm, or counts per minute. In this mode, the 7261A counter can be used as a tachometer. Here the display reads directly in revolutions per minute. In order to achieve this readout, the input must come from a transducer that produces 100 pulses per revolution. To measure the revolutions per minute of a motor shaft or other rotating device, some form of transducer must be connected to it. A magnetic gear-tooth transducer similar to that described in a previous chapter can be used. One having an output of 100 pulses per revolution can then be connected to input channel A. With the counts-per-minute mode selected, the LED display will read the revolutions per minute directly. This is accomplished by having the time base and control logic generate a 600-millisecond gate interval.

Revolutions per minute can also be measured with transducers that have other pulse-per-revolution rates. The display will not read in revolutions per minute directly, but the correct value can be computed by using the displayed value and knowing the number of pulses per revolution of the given transducer.

$$\text{Correct rpm} = \frac{100 \times \text{Displayed rpm}}{N}$$

where N is the number of transducer pulses per revolution.

Frequency-ratio measurements can also be accomplished with the 7261A counter. When this mode is selected, the interval time-base and gating logic is not used. Instead, two external input signals are used. The object is to have the counter measure and display the ratio of two separate input frequencies. These two input frequencies are applied to channels A and B. The higher-frequency signal (5 Hz to 125 MHz) is applied to channel A. The signal on channel B gates the signal on channel A. The display reads the ratio A/B. If input A is 65.536 kHz and input B is 1.024 kHz, the display will read 64. The resolution of the measurement can be improved by having the channel-B input signal divided by the decade time-base dividers. In the frequency-ratio mode, the resolution switch selects one of six divide ratios.

Period measurement is accomplished by applying the input signal to channel A. The period of frequencies in the range 5 Hz through 2 MHz may be measured in this mode. Recall that in the period mode, it is the period of the input signal that determines the main-gate enable duration. During the period interval, the counter accumulates accurate time-base signals. In this case, time-base signals from 1 kHz to 100 MHz in decade steps are counted, depending on the setting of the resolution switch. A maximum resolution of 10 nanoseconds is possible with this arrangement. When the 10-ns range is selected, the counter counts 100-MHz pulses from the 10–100-MHz pll multiplier circuit.

In the period-average mode, the time-base signals are counted for a period of time equal to some decade multiple of the input signal period. The input signal on channel A is decade divided by the time-base dividers and thereby causes the control logic to generate a gate-enable signal some decade multiple ($10^0$ to $10^5$) of the input period. A time-base signal of 1 kHz to 100 MHz in decade steps is selected by the resolution control to determine the resolution. Period-average measurements may be accomplished on signals between 5 Hz and 1 MHz. Because of the increased counting time, the resolution of the measurement is significantly improved. A resolution of 10 picoseconds is possible in this mode because of the averaging action.

To make time-interval measurements, time-base pulses are counted for the period of time during which the main gate is enabled. The main gate is turned on and off by input control signals applied to channels A and B. For example, the time-interval measurement may be initiated by an input to channel A and terminated by an input pulse to channel B.

Recall also that channels A and B may be connected together so that a single input can control the start and stop of the timing function. In this way, characteristics such as pulse width, rise time, and fall time can be measured. By adjusting the trigger levels and slopes of the two input channels, a variety of time-interval measurements can be made on a single signal.

Time-interval averaging measurements can also be made on the 7261A counter. In this mode, the counter displays the average number of counts of the 100-MHz clock that occur during each period of the input signal. The main counter accumulates pulses for N times the input-signal duration (N is some decade value from $10^0$ through $10^5$). The total number of periods averaged is determined by the front-panel resolution-control setting. The time-interval averaging mode, like the period-averaging mode, provides greater resolution and improvement of accuracy.

Another operating mode of the 7261A is the totalize mode, in which the counter is simply used to tally events applied to input channel A. The time-base and control-logic signals that ordinarily operate the main gate are inhibited at this time. In this way, the counter is simply used to accumulate the occurrence of external events on channel A. A special provision of this mode is that the events occurring on channel A may be gated by an input signal applied to channel B. In other words, channels A and B are logically ANDed. In this way, an external gating signal may be used to control the application of events on channel A. Otherwise, channel B may be permanently enabled so that only the channel A totalize function is operational.

The final remaining mode is the self-check function. This mode is used to make a dynamic check of the operation of the counter. It is used to verify the proper operation of all the digital circuitry in the counter except for the input-signal conditioning circuitry and the time-base accuracy. Depressing the front-panel reset button in the self-check mode causes the display to show all 8s. This verifies the operation of all display digits. Setting the resolution control to each of its positions provides for the display of a specific number, thereby indicating the operation of the time base and all the related circuitry.

As indicated earlier, the 7261A counter has two interface options. The first option is a simple monitor and control interface that allows the counter to be controlled by some external source or permits counter measurements to be displayed or recorded externally. For example, the unit may be used with an external input control panel, or the counter output may be sent to an accessory printer. The other interface is a standard IEEE 488 interface that allows the counter to become part of a larger, more complete automated test system.

The simple data input/output interface consists of circuitry that is used to control the function, resolution, and slope of the counter from an external source. Display data from the counter may be transmitted to an external accessory unit such as a printer. Binary-coded decimal data from the display is sent bit-parallel, digit-serial through the interface to the accessory.

The IEEE 488 interface contains an 8048 eight-bit microprocessor. This is a single-chip eight-bit microcomputer containing both RAM and ROM. It is used in conjunction with the 68488 LSI interface chip that implements all of the functions of the IEEE 488 interface bus. With this interface, the function, resolution, and slope of the counter may be set up and controlled from an external source such as a computer. Data may be transmitted to or from the counter via the interface.

CHAPTER 9

# Counter Circuits and Applications

In this chapter, we examine more closely some of the counting circuits used in the instruments described elsewhere in this book. No handbook on digital counters would be complete without a review of the important counter circuits and a look at some of the ways they are used. While this chapter is not a complete reference on the subject, many popular MSI and LSI counter circuits and their applications are summarized.

The main circuit in every counter is the flip-flop. The operation of a flip-flop was reviewed in Chapter 1. Many different counting circuits can be made by simply varying the number and interconnection of the flip-flops. Additional logic circuitry can be added to enhance their performance or create useful features. But, while an almost infinite variety of counters can be created, it turns out that binary and bcd counters are the most widely used. These circuits count input pulses and store the value as a binary or bcd number. Binary and bcd counters were also covered in Chapter 1. Most of the counter circuits discussed here are some form of binary or bcd counter. The main purpose of this chapter is to look at the various types available, their variations, and their applications.

## MSI COUNTERS

While any kind of counter may be constructed from individual integrated-circuit flip-flops and logic gates, this approach is simply not necessary. Most of the popular types of counters are already available as complete integrated circuits. These are usually described by the phrase *medium-scale integration* (MSI). Each counter is made on a single silicon chip, and all the flip-flops and related logic gates are already interconnected to perform a specific counting function.

The most common MSI counters in use are standard binary and bcd types. The binary counters are usually four-bit versions. Some eight-bit binary counters are also available. To make longer counters, four-bit MSI binary counters are simply cascaded.

Three basic types of MSI counters are available: TTL, ECL, and CMOS. By far the most widely used are the TTL (transistor-transistor logic) counters, which can be obtained in both bcd and four-bit binary types. The average TTL counter has a maximum counting rate of approximately 30 MHz. Low-power versions of these counters have a maximum count rate of 3 MHz. High-speed versions can count as high as 50 to 125 MHz. The most popular TTL counters are the low-power Schottky versions. These feature the low power consumption of a standard TTL counter but a counting speed of 25 to 40 MHz.

For counting applications that exceed the capabilities of a TTL counter, ECL (emitter-coupled logic) counters can be used. Virtually all ECL counters can count beyond 100 MHz. For example, 500-MHz ECL counters are common. The upper counting rate for modern ECL counters is approximately 2 GHz. Both bcd and four-bit binary versions can be obtained. Such counters are widely used as prescaling frequency dividers in high-frequency counters.

Over the past several years, MSI counters using CMOS integrated-circuit technology have become more widely used. While their maximum counting rate is limited, their very low power consumption makes them desirable in remote, portable, battery-operated applications. The upper count rate for most CMOS counters is approximately 8 to 10 MHz. A few special CMOS counters can count as high as 20 to 25 MHz. Four-bit binary and bcd counters are typical, but some 7-, 12-, and 14-bit CMOS binary counters are available.

Other MSI counter variations exist. For example, dual four-bit binary or dual bcd counters are available. Some MSI counters also contain storage latches and decoder/drivers for external displays. Such devices greatly reduce the number of ICs required to implement a given digital function.

## SYNCHRONOUS VERSUS ASYNCHRONOUS

There are two major classifications of digital counters in use: the *synchronous*, or *clocked*, *counter* and the *asynchronous*, or *ripple*, *counter*. The ripple, or asynchronous, counter is the simpler of the two.

### Ripple Counters

A ripple counter is illustrated in Fig. 9-1. This four-bit binary counter is made up of cascaded JK flip-flops. The output of one flip-flop is connected to the clock input of the next in sequence. The input signal to be counted is applied to the first flip-flop. Each time an input pulse arrives, the first flip-flop is toggled, or complemented. Typically, these flip-flops change state on the negative-going transition of the clock (C) input signal. The second flip-flop is, in turn, toggled by the first. The third flip-flop is toggled by the second, and so on. Keep in mind that each flip-flop has a finite amount of propagation delay. This delay is the time that it takes for the input transition to be recognized, the circuit to respond, and the output state to change. This delay is only a few nanoseconds in high-speed TTL or ECL flip-flops, but it can be as long as one-half microsecond in lower-speed flip-flops of the CMOS variety. Since one flip-flop triggers the next, the count *ripples* through the counter. The propagation delays are additive. For example, if the flip-flops in Fig. 9-1 each have a 50-nanosecond propagation delay, the total delay from the input transition to the output transition of the last flip-flop is 200 nanoseconds.

If a great number of flip-flop stages is used and the propagation delay for each is long, it is possible for the counter to make counting errors. In other words, because of the cumulative propagation delay, the binary number contained within the counter may be incorrect for a short period of time after the input pulse occurs. If the input frequency is high enough, the time between the incoming pulses will be less than the total propagation delay of the counter. When this condition exists, invalid binary code states will exist continuously because the counter never "catches up." In the example given earlier, the four-bit counter in Fig. 9-1 has a total delay of 200 nanoseconds. This translates to a frequency of $1/(200 \times 10^{-9}) = 5$ MHz. In practice, this counter should not be used in any application in which the anticipated counting rate would be likely to exceed that value.

In addition to the invalid states that can occur in a ripple counter at high frequencies, another problem shows up in some applications. When the counter states are decoded, the propagation delays in the flip-flops can cause decoding spikes, or "glitches," to appear at the output of the decoding gates. Refer to Fig. 9-1. In the decoding of state $A\bar{B}\bar{C}D$, glitches occur. These glitches are generated by the momentary invalid binary states in the counter that exist because of the propagation delay. Decoding spikes may cause false triggering or improper operation of the other circuits to which they are applied.

## Synchronous Counters

The problems of decoding spikes and invalid counting states in a ripple counter can be overcome by using a synchronous counter. In a synchronous counter, the input signal is applied to all flip-flop clock inputs. To determine the proper count sequence, the various flip-flop outputs are connected to the J and K inputs of the succeeding flip-flops with AND gates. The four-stage synchronous counter shown in Fig. 9-2 illustrates this. With this arrangement, all flip-flops change state simultaneously upon the arrival of the input clock signal. But, the state change of each flip-flop depends on the states of the preceding stages.

This circuit is not without propagation-delay effects, however. While all flip-flops are clocked simultaneously, they will not change state unless the proper conditions exist at the J and K inputs. This means that the flip-flop outputs must propagate through any logic gates connected between the flip-flop ouputs and the J and K inputs. The total propagation delay, however, is only that of one flip-flop and one AND gate. As a result, the potential count rate is significantly higher than that of a ripple counter of the same length. If the propagation delay of the flip-flops is 50 nanoseconds and the delay of an AND gate is 10 nanoseconds, the total delay is 60 nanoseconds. This translates to an upper count limit of $1/(60 \times 10^{-9}) = 16.67$ MHz.

Synchronous counters offer the advantage of much higher counting speed. Since propagation delays are not additive, a higher count rate can be accommodated, and no count errors or invalid states will exist. In addition, decoding spikes are not generated. The outputs of decoder gates connected to a synchronous counter are essentially glitch-free. Because of their higher counting speed and glitch-free operation, synchronous counters are certainly to be preferred. However, they are more complex and consume more power because of the extra circuitry usually involved.

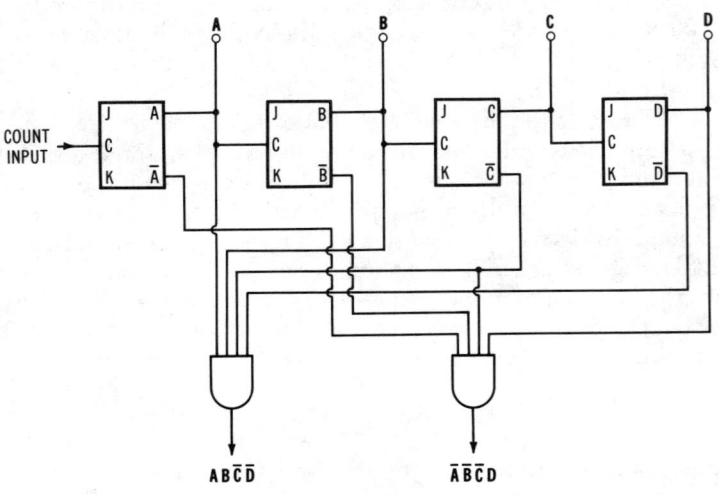

(A) *Logic diagram.*

Fig. 9-1. Asynchronous

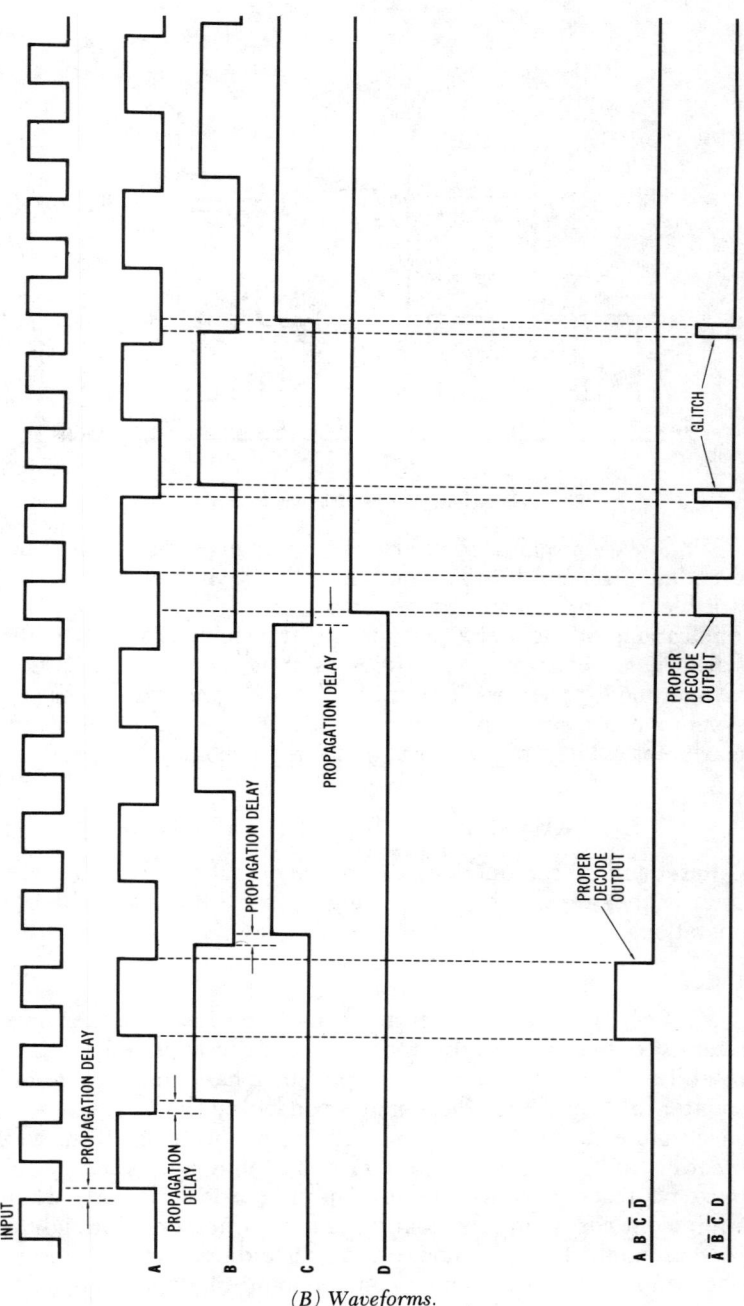

(B) Waveforms.

(ripple) counter.

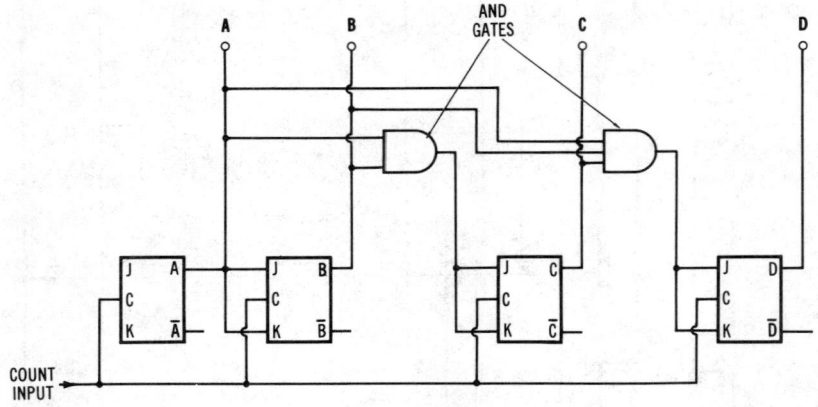

Fig. 9-2. A four-bit synchronous binary counter.

While synchronous counters solve the high-speed counting and decoding problem, this does not mean that asynchronous counters are worthless. Ripple counters perform more than adequately in many applications. While their count speed may be more limited than that of a synchronous counter, in most cases it is usually sufficient. Many asynchronous counters are available and are widely used. The synchronous counter is preferred only in those applications where maximum counting speed and glitch-free decoding are necessary.

## RESET AND PRESET FUNCTIONS

Reset and preset functions are special operations used to initialize a counter before it begins counting. Most counters include these operations.

### Reset

Most counters have a reset input line. To reset a counter means to clear its contents to zero. In other words, all flip-flops in the counter are reset. Resetting is the same as entering the binary number zero into the counter flip-flops. Regardless of how you look at the reset function, it is an absolute necessity in most counting applications. If an accurate record is to be kept of the number of incoming pulses, every counter must be reset prior to its use. This is particularly true when the equipment containing the counter is first turned on. Normally, when power is applied, the flip-flops come up in either their set or reset state. The counter, therefore, may contain some random number. If this random number were left in the counter prior to the counting operation, the resulting count would be in error. For that reason, most counters are reset prior to beginning their operation. This reset function is performed by a logic signal applied to the input pin of the counter.

**Preset**

Presetting means entering some specific value into the counter prior to beginning the count operation. A preset counter is one in which any desired binary value may be loaded before a count operation is performed. When a counter is being preset, it acts very much like a simple storage register. Typically, the binary number is applied in parallel to the counter through external input lines, one input bit per counter flip-flop. A four-bit preset counter, for example, has four separate input lines through which a four-bit binary number may be entered. The desired input is loaded by enabling or triggering the preset circuitry with an external signal. The counter to be preset can receive its inputs from another counter, a shift register, a storage register, a thumbwheel switch, a set of toggle switches, or any other binary-number source.

## UP/DOWN COUNTERS

Most of the counters we have discussed are up counters in that as each input pulse occurs, the count is incremented. That is, the binary number stored in the counter increases by one to reflect each new input count.

There is also such a thing as a down counter. With only minor modifications to the way the flip-flops are interconnected, a down counter can be easily created. When input pulses occur, the count is decremented. For each input pulse, the binary number stored in the counter is reduced by one. Like up counters, down counters find wide application in digital equipment.

Because the circuitry for up and down counters is so similar, it is relatively easy to combine both the up and down count functions within the same counter. While many up counters are available in integrated-circuit form, no down counters are available. Instead, counters which can count up and down are more typical. A number of different versions are available. One version contains a single input line to which the pulses to be counted are applied. A special control line is used to control whether the count is up or down. In another version, two count input lines are provided. Pulses to increment the counter are applied to one input, and pulses to decrement the counter are applied to the other input. With this arrangement, two separate pulse sources may be used. A single set of counting flip-flops will be alternately incremented and decremented by the incoming pulses.

Fig. 9-3 shows a logic diagram of a four-bit synchronous MSI binary up/down counter. Note the two separate inputs for down count and up

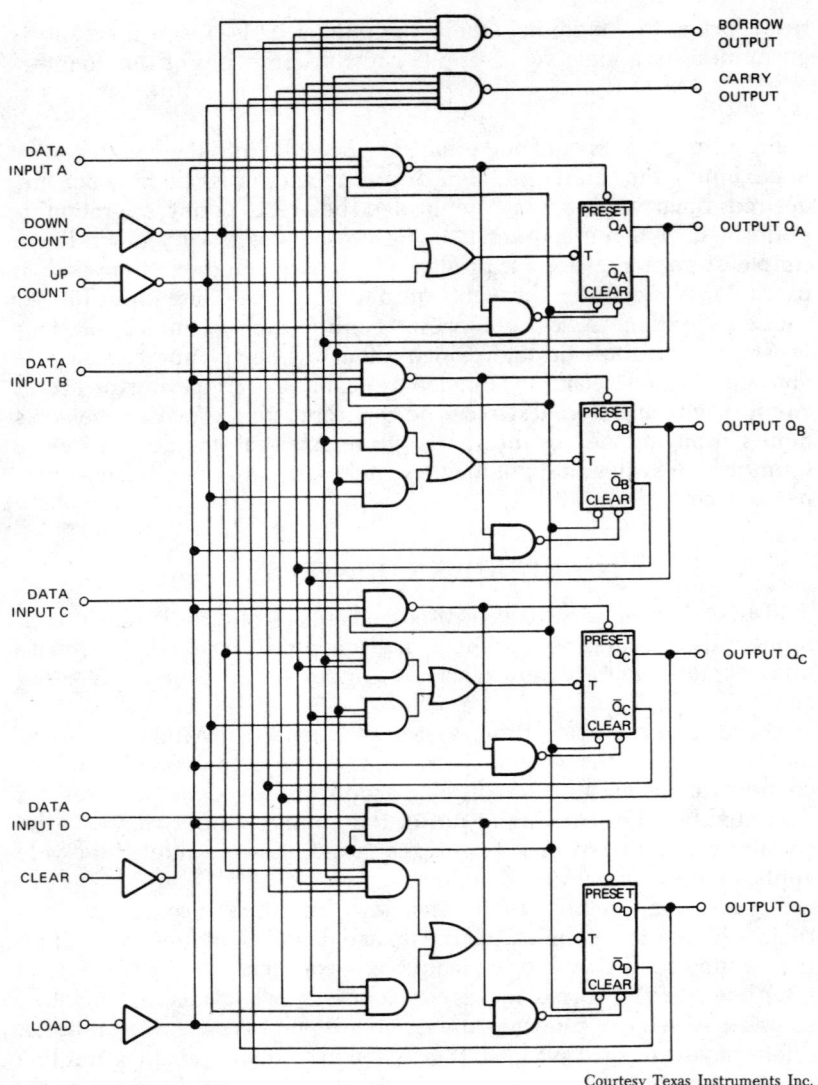

Courtesy Texas Instruments Inc.

Fig. 9-3. A four-bit binary synchronous MSI up/down counter.

count. The counter also has a clear input, which is used to reset the counter. This counter may also be preset. The four-bit number to be entered is applied to the data input A, B, C, and D lines. The flip-flops are then preset when a logic 0 occurs on the load input. The borrow and carry outputs are used for cascading these counters. The borrow output feeds the down-count input on the next counter in sequence, and the carry output feeds the up-count input on the next counter in sequence. A bcd version of this counter is also available.

You have already seen one basic application of up/down counters in a previous chapter. This was the use of an up/down counter in industrial control applications. The counter was used primarily as a totalizer in which each input pulse caused an additional count to be added. However, under special conditions, it was necessary to decrement the counter, or subtract counts. Up/down counters are ideal for such simple add and subtract functions.

## COUNTERS AS DIVIDERS

Counters can also be used in frequency-division applications. A flip-flop is basically a divide-by-2 frequency-divider circuit. The output of a flip-flop is half the frequency of the input. When flip-flops are cascaded, other values of frequency division can be performed. With a binary counter, the frequency division is some power of 2, as expressed by the relationship $2^N$, where N is the number of flip-flops used. A four-bit MSI counter, for example, will provide frequency division by $2^4$ = 16. This is true of both up and down counters.

A bcd counter can also be used as a frequency divider. As you have already seen, a bcd, or decade, counter produces frequency division by 10. Flip-flops can be interconnected in any number of ways to perform frequency division by any integer value. Usually, these integer values can be readily obtained by mixing and cascading available binary and/or bcd counters.

One type of bcd counter consists of separate divide-by-2 (a flip-flop) and divide-by-5 sections. When interconnected, they produce 8-4-2-1 bcd-code counting and division by 10. However, the device is designed so that the divide-by-5 and divide-by-2 sections may be electrically separated and used independently. The popular 7490 series TTL counters have this configuration.

Another popular TTL MSI counter is the 7492, which consists of a single flip-flop that divides by 2 and a three-bit section that divides by 6. When interconnected, they divide by 12. These two sections can be used independently. In fact, the divide-by-6 section actually consists of a two–flip-flop circuit that divides by 3 and a single divide-by-2 flip-flop. The divide-by-3 section may be used independently.

Several special MSI counters are available that can be programmed to perform frequency division by any integer value. One four-bit TTL device is programmed by entering the desired value in binary form. The unit is then set up so that it performs frequency division for any integer value between 2 and 16.

## APPLICATIONS OF BINARY AND BCD COUNTERS

Counters are so widely used in digital equipment that it is almost impossible to cover all their possible uses. You have already seen how bcd counters are used in test instruments to perform time and frequency measurements. Here we would like to give you several additional examples of the use of counters in a variety of digital applications. The emphasis is on MSI binary and bcd counters, since they are the most widely used.

### Frequency Dividers

As indicated earlier, MSI counters can be used as frequency dividers. The various types can be cascaded to generate frequency division by any integer value. Two examples are shown in Fig. 9-4.

In Fig. 9-4A, a total division of 800 is required. This is obtained by cascading dividers of 16, 5, and 10 (16 × 5 × 10 = 800). The ÷16 divider is simply a single four-bit MSI binary counter. The ÷5 section can be obtained with the three-bit ÷5 section in a 7490 counter. The ÷10 stage is a bcd counter. Note that with an 8-MHz input, the output is 8,000,000/800 = 10,000 Hz, or 10 kHz.

In Fig. 9-4B, a division of 144 is obtained. The ÷6 and ÷3 functions can be obtained with the appropriate circuits in the 7492 MSI counter described earlier. The ÷8 function can be obtained with a four-bit binary counter; only the first three flip-flops are needed to give division by 8. The input of 288 kHz is divided by 144 to obtain a 2-kHz output. In both examples, intermediate output frequencies are also available.

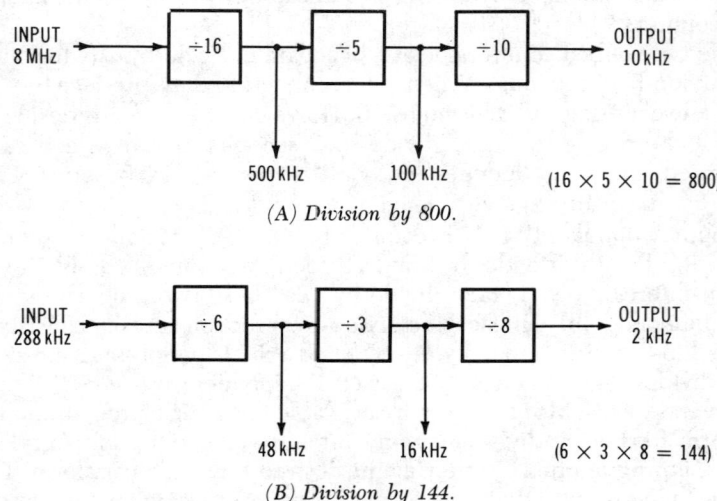

Fig. 9-4. Examples of frequency division with MSI counters.

## Binary/BCD Conversions

In some digital applications, it is necessary to convert a binary number into its bcd equivalent or to convert a bcd number into an equivalent binary number. These conversions can be performed by special combinational logic circuits (gate arrays). However, counters can also be used.

Fig. 9-5 shows an arrangement used to convert any eight-bit binary number into its bcd equivalent. With eight bits, numbers from 0 (00000000) to 255 (11111111) can be represented. With a three-digit output possible, three bcd counters are needed.

Assume that we want to convert the binary number 11001111 (decimal 207) into bcd form. This binary number is loaded into a presettable binary down counter. Two four-bit MSI presettable up/down counters set for down counting are used. The down counter is decremented by clock pulses from NAND gate 1. The clock pulses also increment the three-digit bcd counter.

To prepare the circuit for proper operation, a signal is applied to the "initialize" input. This loads the binary number into the down counter and resets the bcd counters to zero. Next, a start pulse is applied to the S input of the control flip-flop. This sets the flip-flop, and its Q output goes to a logic 1, thereby enabling gate 1. The clock pulses thus simultaneously cause the bcd counter to be incremented and the binary counter to be decremented.

The inverters at the outputs of the binary counter and NAND gate 2 are used to detect the zero (00000000) state in the down counter. After 207 clock pulses, the down counter will be decremented to zero. When this occurs, the output of NAND gate 2 goes to logic 0, thereby resetting the control flip-flop. This inhibits NAND gate 1 so that no more clock pulses reach the bcd counters. At this time, the bcd counters contain the bcd equivalent of 207, which is 0010 0000 0111. The speed of the conversion is a function of the clock frequency.

This technique can also be used to perform bcd-to-binary conversions. The positions of the binary and bcd counters in Fig. 9-5 are simply interchanged. The bcd number to be converted is loaded into a bcd down counter, which is then decremented to zero as a binary up counter is being incremented.

## Address Generation

In computers, mass-storage peripheral devices such as magnetic tapes and disks are used as auxiliary storage media. The data to be stored is magnetically recorded on the tape or disk. The data can be retrieved by reading it from the tape or disk.

The data to be stored is usually serialized, that is, recorded sequentially a bit at a time. Most computer data is in the form of *bytes*, eight-bit words. A certain number of bytes are grouped together to form

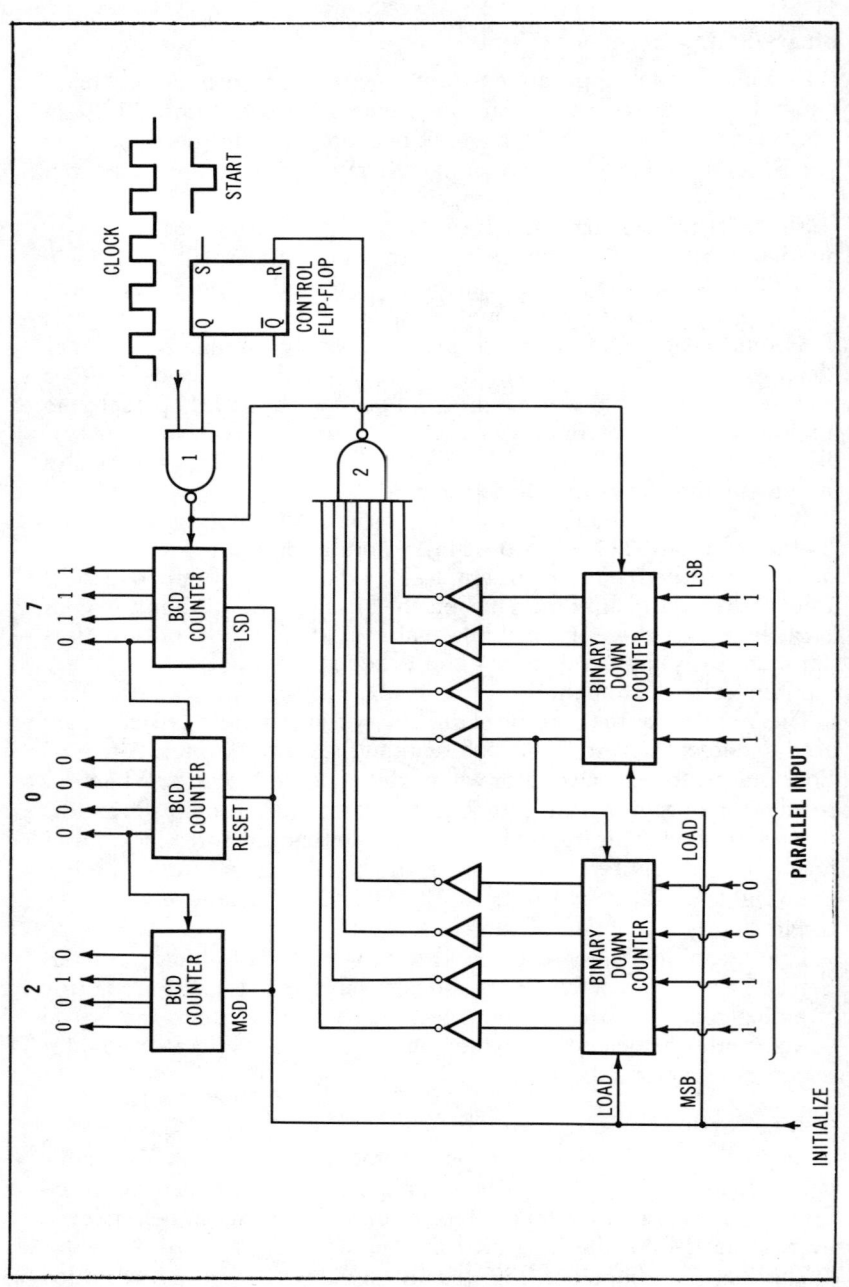

Fig. 9-5. A converter for changing binary code to bcd code.

*blocks*, or *sectors*, of data. In order to locate a specific byte in a given sector, an addressing scheme is used.

Fig. 9-6 shows the format in a hypothetical floppy-disk storage system. The disk is set up to store data on 32 concentric tracks. Each track is given a number, or address, so that it can be identified. Assume that the tracks are numbered from 0 to 31. Next, the disk, and as a result each track, is divided into 16 sectors. These sectors are numbered 0 through 15. Each track in each sector can store 128 bytes of data. The bytes in each sector are also numbered, in this case from 0 to 127. Finally, remember that each byte contains eight bits. With this arrangement, the total storage capacity of the disk is 128 bytes × 16 sectors × 32 tracks = 65,536 bytes (64K).

Disks are generally regarded as random-access storage devices in that any given sector can be located. Since each byte in each sector on each track is numbered, then theoretically it can be found. These numbers are referred to as *addresses*. Binary counters are used to generate and keep track of these addresses. One arrangement is shown in Fig. 9-7.

When the binary data is recorded, it is transmitted serially (a bit at a time) to the magnetic recording head. This serial transmission is usually accomplished by a shift register stepped by a clock signal whose

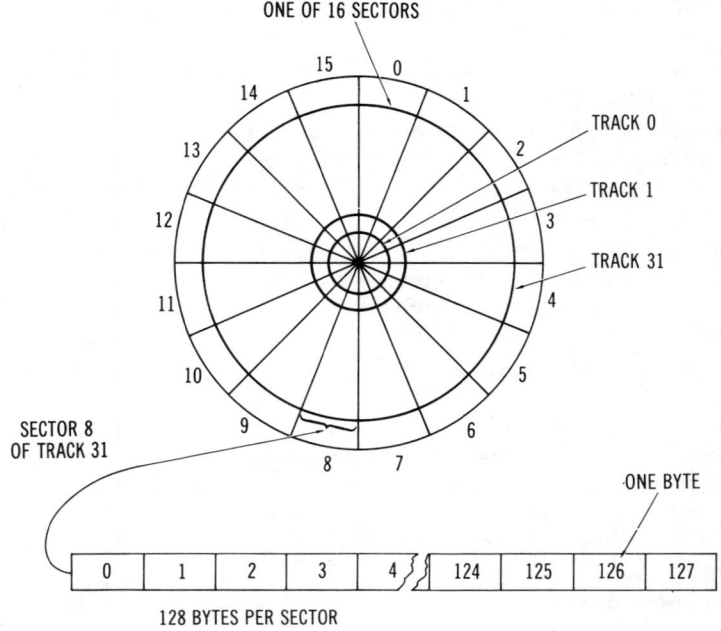

Fig. 9-6. Data-storage format on a hypothetical floppy disk.

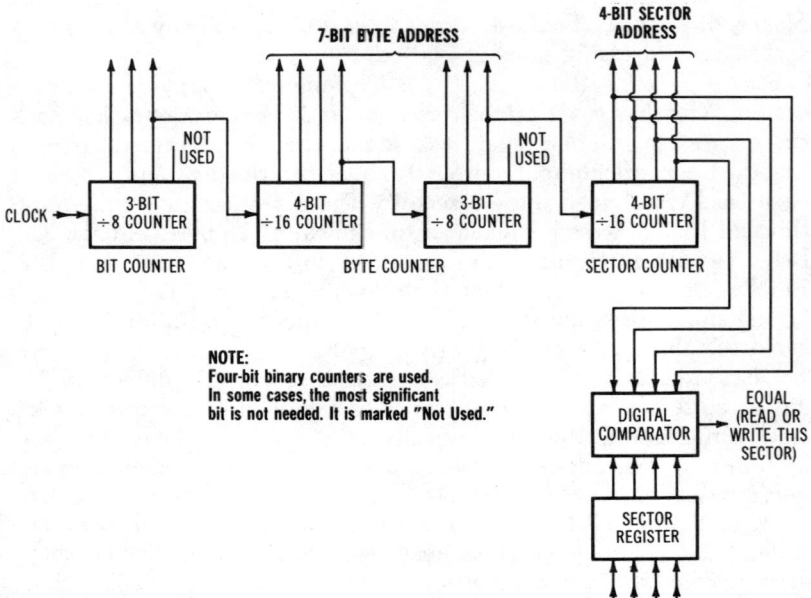

Fig. 9-7. Binary counters used as address generators for a floppy-disk storage system.

frequency controls the speed. This clock is also synchronized with the rotation of the disk. A hole in the disk at the beginning of sector 0 is often used as a reference to tell the circuitry when to start writing (recording) or reading the data. The clock also steps a three-bit binary counter that counts the bits of each word (byte) to be recorded. Each time eight bits are counted, the three-bit counter recycles and increments a seven-bit byte-address counter. Every eight clock pulses, one byte is recorded or read out. The seven-bit byte counter is stepped after each eight-bit count cycle. For example, the first byte in a sector is designated as the 0 byte. Its address in the seven-bit counter is 0000000. After this first byte is stored or read out as indicated by counting eight clock pulses, the seven-bit byte counter is incremented to 0000001, the address of the next byte in sequence.

The seven-bit byte counter, in turn, steps a four-bit sector counter. This counter contains the sector addresses, 0 (0000) through 15 (1111). When the seven-bit byte counter cycles from 1111111 to 0000000, it steps the sector counter, indicating that one 128-byte sector has been counted.

In most disk storage systems, individual bytes of data are not addressed. In practice, only specific sectors on indicated tracks are located. The data in a sector is then read out sequentially as a block. To find a given sector, a digital comparator is used. See Fig. 9-7. The desired sector number is loaded into the sector register. This sector address is applied to the comparator. The clock then steps the counters until the address in the sector counter equals the address in the sector register. At this time, the comparator generates an "equal" output signal that tells the circuitry either to record or to read data.

## Digital Frequency Display

Another good example of a counter application is the use of frequency counters built into stereo tuners and communications receivers. Such counters are virtually identical to the test instruments discussed elsewhere in this book. However, because of their small size and low cost, counters can readily be built into receiving equipment to allow direct digital display of the frequency being received. Such a display usually replaces the ordinary pointer-type tuning dial.

In this application, the frequency counter cannot be used to measure the frequency of the incoming signal directly. The signal is far too low in amplitude at the receiver input to permit reliable counting. However, this is not a disadvantage. The actual received value can be readily determined by other means. The most common technique for determining the frequency of the incoming signal is to measure the frequency of the local oscillator. Such an arrangement is shown in Fig. 9-8.

Recall that most receivers are superheterodynes, which mix the incoming signal ($f_s$) with a local-oscillator signal ($f_o$) to create a new lower-frequency signal referred to as the intermediate frequency, or if (also designated $f_i$ in Fig. 9-8). The if is a fixed, predetermined value. In most am receivers the if is 455 kHz. In most fm receivers, the if is 10.7 MHz. Usually, the local-oscillator frequency is higher than the incoming frequency by the value of the if:

$$f_i = f_o - f_s \quad \text{or} \quad f_o = f_s + f_i$$

If the frequency counter is connected to the local oscillator, the frequency display will be higher than the incoming signal by the amount of the if. For that reason, the if value must be subtracted from the local-oscillator frequency to get the correct input-signal frequency. The easiest way to accomplish this is with a preset counter. The main bcd counter in Fig. 9-8 is preset with a value that will cause the correct incoming frequency to be displayed when the local-oscillator frequency is measured.

The number to be preset into the counter under these conditions is the tens complement of the if value. The tens complement is simply $10^N - f_i$, where N is the number of digits in the display and $f_i$ is expressed in the same units as the display. The am receiver in Fig. 9-8 has a four-digit frequency counter that reads in kilohertz (kHz). The if value is 455 kHz.

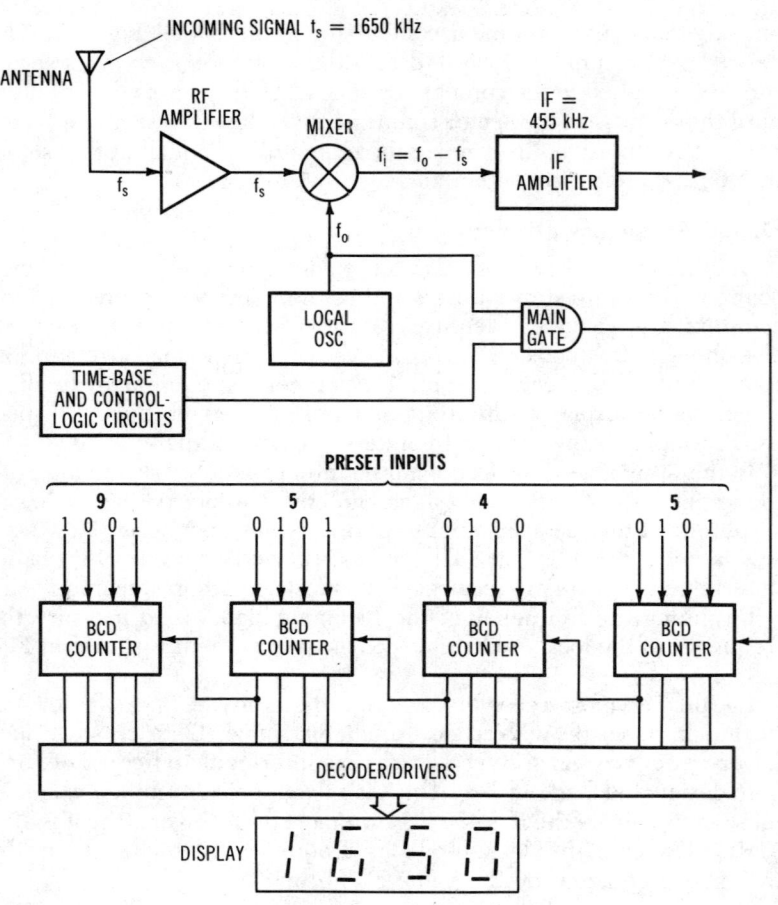

Fig. 9-8. Use of a preset counter to measure and display the received frequency in a superheterodyne receiver.

To find the tens complement, you simply subtract 455 from $10^4$. The tens complement, therefore, is $10,000 - 455 = 9545$. This is the number which is preset into the bcd counters prior to the measurement of the local-oscillator frequency. With an incoming signal at 1650 kHz, the local oscillator is tuned to $1650 + 455 = 2105$ kHz. With the content of the counter initially at 9545, the first 455 input pulses will cause the counter to be incremented to 9999 and then recycled to 0000, or be reset. Thereafter, the counter will count the remaining number of pulses (1650) equal to the input frequency. After each measurement cycle, the counters are again preset.

# TIMERS

Timers are special MSI counter circuits that are designed primarily for generating accurate timing pulses. A timer incorporates a counter that accumulates pulses from an internal oscillator to generate an output pulse of some predetermined value. Timers can generate accurate output pulse durations from microseconds to days. They can also be used to generate time delays. By cascading several timers, a series of accurately sequenced timing pulses can be generated for a variety of special-purpose control applications.

Fig. 9-9 shows diagrams of the XR-2242 timer. This timer consists of a time-base oscillator, a control flip-flop, and an eight-bit binary counter.

The frequency of the internal time-base oscillator is set by an external RC network. The resistor and capacitor values determine the time-base pulse period, T, which is equal to $R \times C$. For example, with a 1-$\mu$F capacitor and a 1-megohm resistor, the pulse interval will be $1 \times 10^{-6} \times 1 \times 10^{6} = 1$ second. In other words, the time-base oscillator will generate a 1-Hz signal.

A timing cycle is initiated by applying a pulse to the trigger input of the control flip-flop. This sets the flip-flop and enables the time-base oscillator. The eight-bit binary counter then begins to count the time-base pulses. Note that only the outputs of the first and eighth flip-flops in the binary counter are made available. These flip-flop outputs drive open-collector transistors.

After 128 clock pulses have been accumulated, the output of flip-flop 8 changes state (Fig. 9-9C). Normally it is the output of this last flip-flop that is connected back to the reset input of the control flip-flop. Thus after 128 clock pulses have occurred, the control flip-flop is reset, thereby stopping the count. The total length of the timing pulse then is 128RC. With the RC network mentioned earlier, the output pulse duration would be 128 seconds, or 2 minutes and 8 seconds. Very short timing pulses can be created by making the resistor and capacitor values smaller. On the other hand, making these values large will create ultralong time intervals. With one of these circuits, a pulse as long as a day can be created. By cascading two of these circuits, time delays of up to one year can be generated.

Other popular timer circuits are the Intersil 8240 and 8250. These devices are similar to the one just described but are somewhat more sophisticated in terms of capabilities. Like the XR-2242, each consists of an internal time-base oscillator, a control flip-flop, and a counter. The 8240 contains a standard eight-bit binary counter, and the 8250 contains a two-digit bcd counter. A general block diagram of these devices is shown in Fig. 9-10.

An important feature of the 8240/50 is the ability to detect a specific counter state. The output from each of the eight counter flip-flops is an

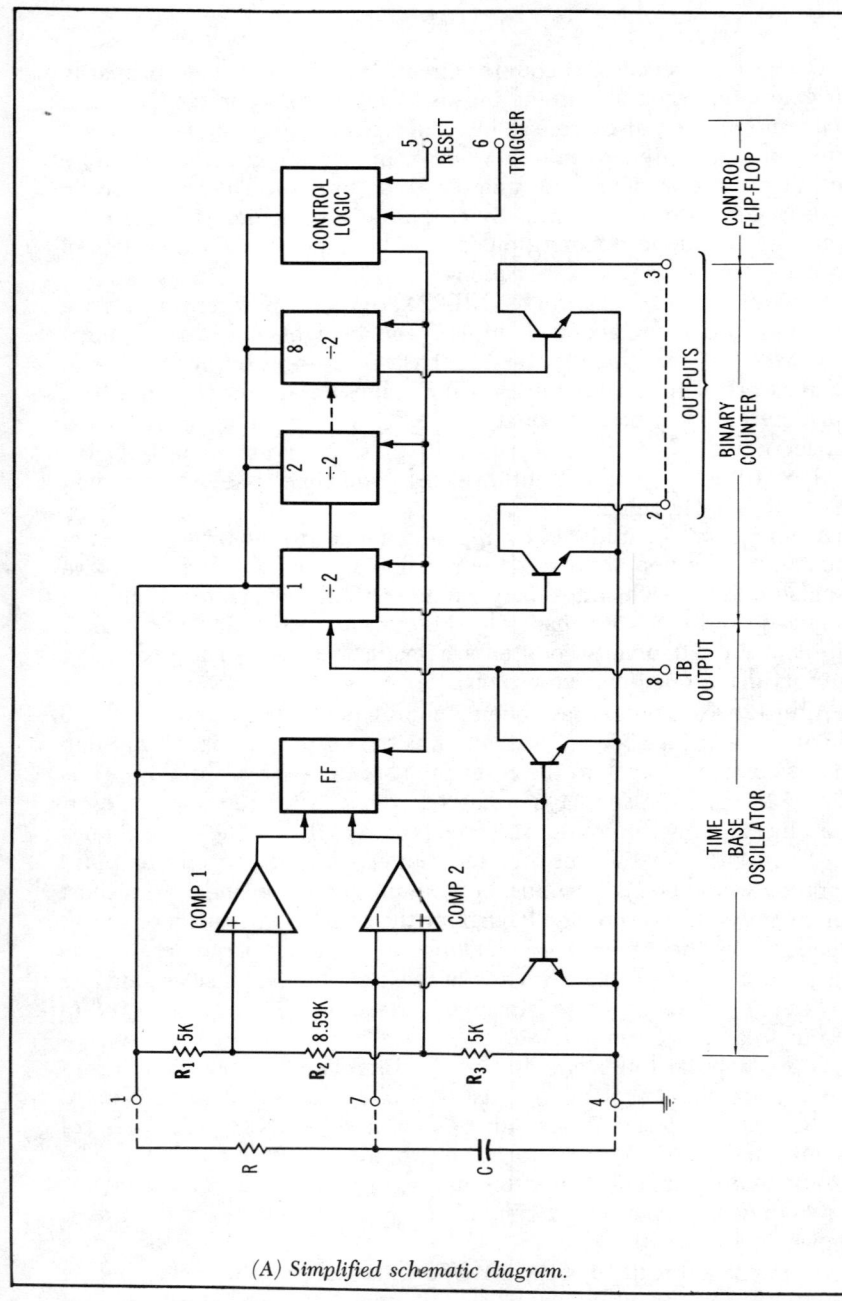

(A) Simplified schematic diagram.

Fig. 9-9. Exar

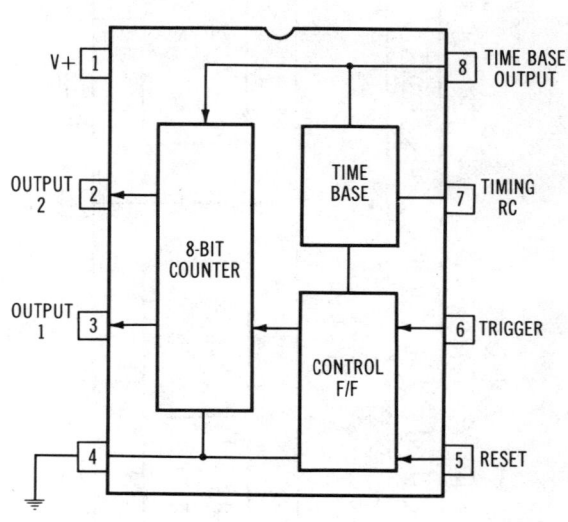

*(B) Functional block diagram.*

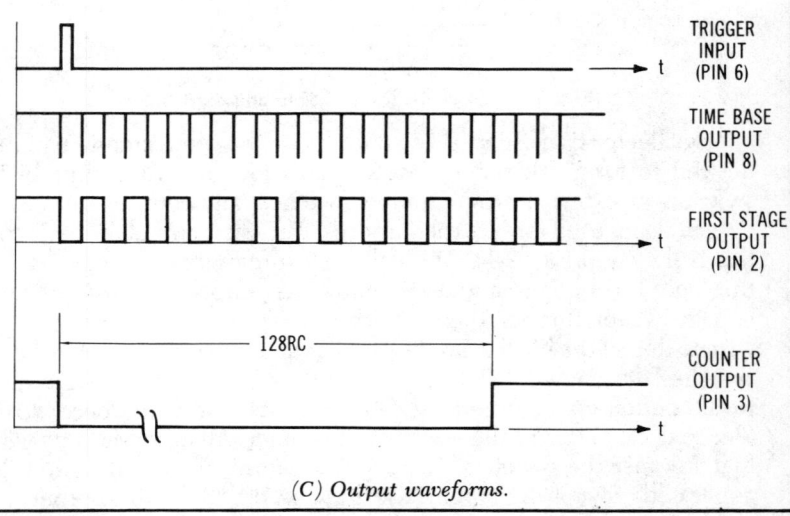

*(C) Output waveforms.*

**XR-2242 timer.**

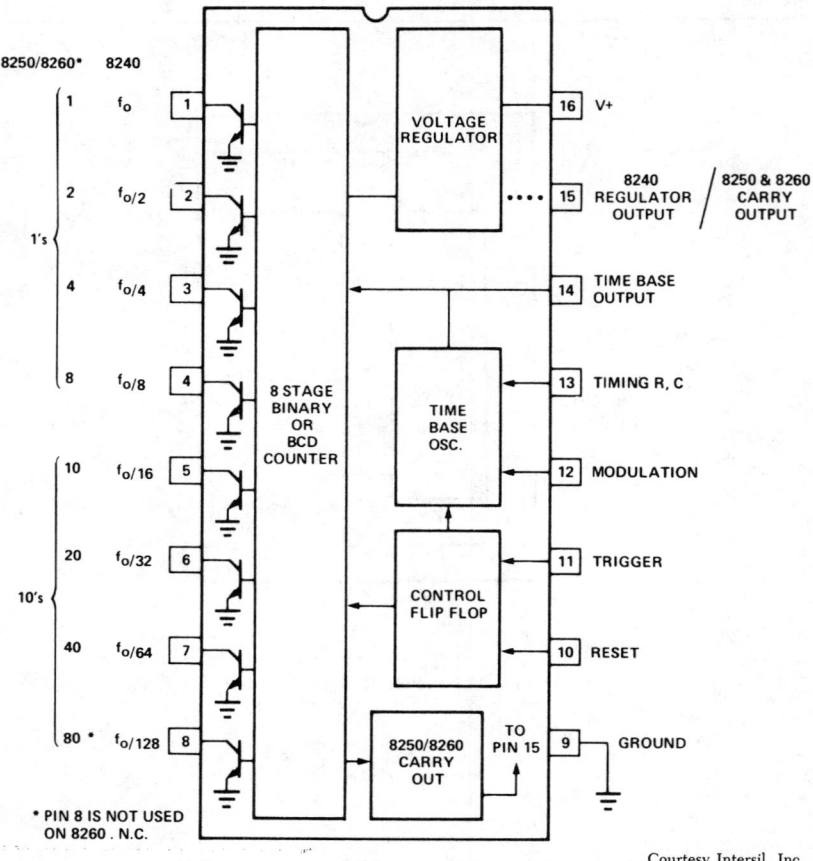

Fig. 9-10. Block diagram of 8240 and 8250 timers.

open-collector transistor. These outputs can be connected in parallel as desired to form a wired logical AND function. By using thumbwheel switches or external programming switches, any counter state can be detected and used for control purposes. The basic circuit for performing this is shown in Fig. 9-11. With all of the programming switches open, the logic level appearing at the output is simply the positive supply voltage as seen through the load resistor, $R_L$. However, assume that we wish to detect the binary number 10001, or the decimal number 17. The switches designated 16T and 1T would be closed. All other switches would be left open. When the counter contains the number 00010001 (decimal 17), outputs 16T and 1T will be high. All other outputs will be high because the switches are open. Therefore, an output pulse will be generated. Switch S1 would also be closed so that the output pulse would be fed back around to the reset input.

To start a timing cycle, a positive trigger pulse is applied to input pin 11. This sets the control flip-flop and enables the time-base oscillator. The time-base oscillator then begins incrementing the counter. When counter state 10001 occurs, the 16T and 1T outputs both go high. All other outputs will be high. This causes the output terminal to go high. A reset pulse will be applied to the control flip-flop. This stops the counting cycle. The output pulse duration in this case is simply 17T, where T is equal to the time constant (RC) of the external resistor and capacitor connected to the time-base oscillator.

All kinds of timing functions can be implemented with these timer/counters. They can be used in industrial applications for sequentially enabling various processes. These timers can also be used in home appliances like washing machines to time cycles of operation. Another application is a darkroom timer. These devices can even be used for frequency synthesis, in which any number of known frequencies can be selected and generated. Such frequency synthesizers can be used in electronic musical instruments. Finally, any type of time-delay operation can be implemented with such timers.

## RATE MULTIPLIER

A *rate multiplier*, sometimes referred to as a *discrete multiplier*, is a special digital counter used for scaling down a pulse stream to some specified fraction. In other words, the rate multiplier is a form of frequency divider. This special counter essentially multiplies the input frequency by a scaling factor that is some number between 0 and 1. If the input frequency is designated $f_1$, the output frequency, $f_0$, can be expressed by the relationship $f_0 = Kf_1$, in which K is some number between 0 and 1. The resolution of the multiplying constant is limited by the size of the rate multiplier, that is, by the number of bits used to implement the counter in the rate multiplier.

The heart of a rate multiplier is a counter. Either a binary or bcd counter may be used. When a binary counter is used, the circuit is usually called a binary rate multiplier. When a bcd counter is used, the circuit is usually referred to as a decade or bcd rate multiplier. Both binary and bcd rate multipliers are available in MSI integrated-circuit form.

Fig. 9-12 shows a logic diagram of the Texas Instruments 7497 synchronous MSI binary rate multiplier. It consists of six cascaded flip-flops and operates in exactly the same way as a standard synchronous binary up counter. With six stages, the counter can produce a maximum frequency division of $2^6 = 64$. In other words, the six-bit binary counter has 64 discrete states representing the binary numbers 000000 (0) to 111111 (63).

The primary difference between a standard binary counter and the rate multiplier is the set of AND gates associated with the flip-flops. These are labeled A through F in Fig. 9-12. The complement output of each flip-flop is applied to an AND gate along with other flip-flop outputs. The AND-gate outputs are then ORed to create the final output, Y. Note that the AND gates have external inputs labeled A through F.

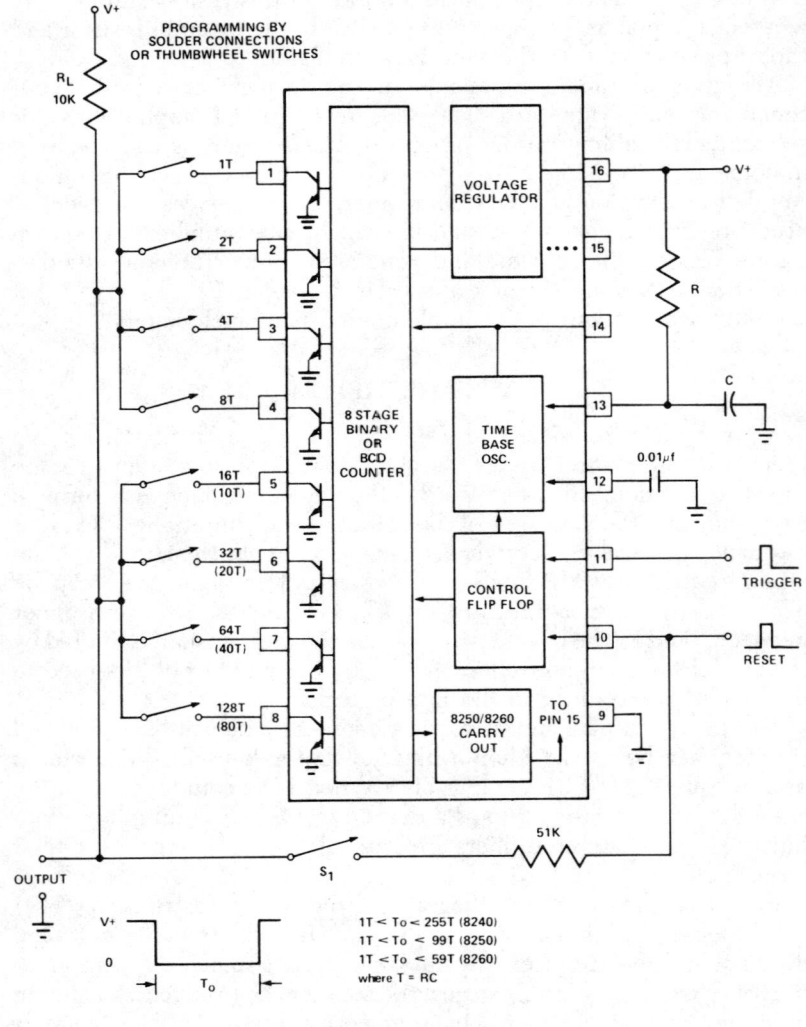

(A) Generalized circuit connection.

Fig. 9-11. Block diagram and timing

A binary number that determines the multiplication factor is supplied to the external input lines on the AND gates. A six-bit binary number is applied to input terminals F through A. The A input is the least significant bit of the number, and the F input is the most significant bit. If all of the binary inputs are zero, the input clock frequency will be multiplied by zero, and no signal will appear at the output. If the binary inputs are set to some value between 000001 and 111111, the input frequency will be multiplied by that number treated as a binary fraction. The output frequency is expressed by the relationship

$$f_o = \frac{M}{64} f_{in}$$

where,

$f_o$ is the output frequency,
$f_{in}$ is the input frequency,
M is the decimal equivalent of the six-bit binary input.

The binary input is treated as a fractional number. If the input is 100001 (decimal 33), the output frequency is

$$f_o = \frac{33}{64} f_{in}$$

We can express 33/64 as a decimal value: 0.515625. If we consider the binary input as a fractional value, or 0.100001, and convert it to its decimal equivalent, we find it to be the same as 33/64, or 0.515625. What this really means is that for every 64 input pulses, 33 output pulses will occur.

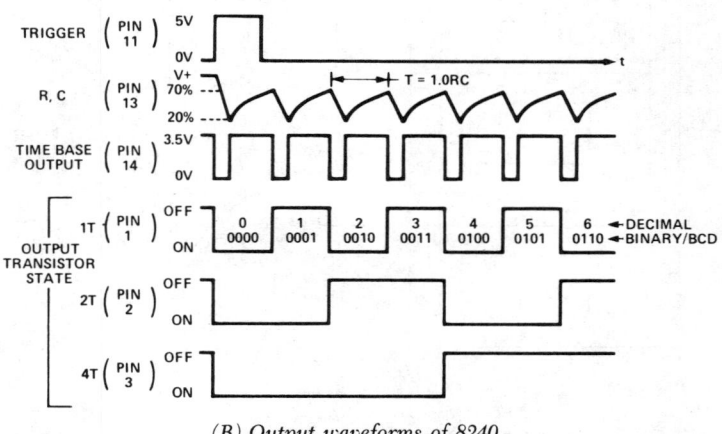

(B) Output waveforms of 8240.

Courtesy Intersil, Inc.

waveforms of 8240 and 8250 timers.

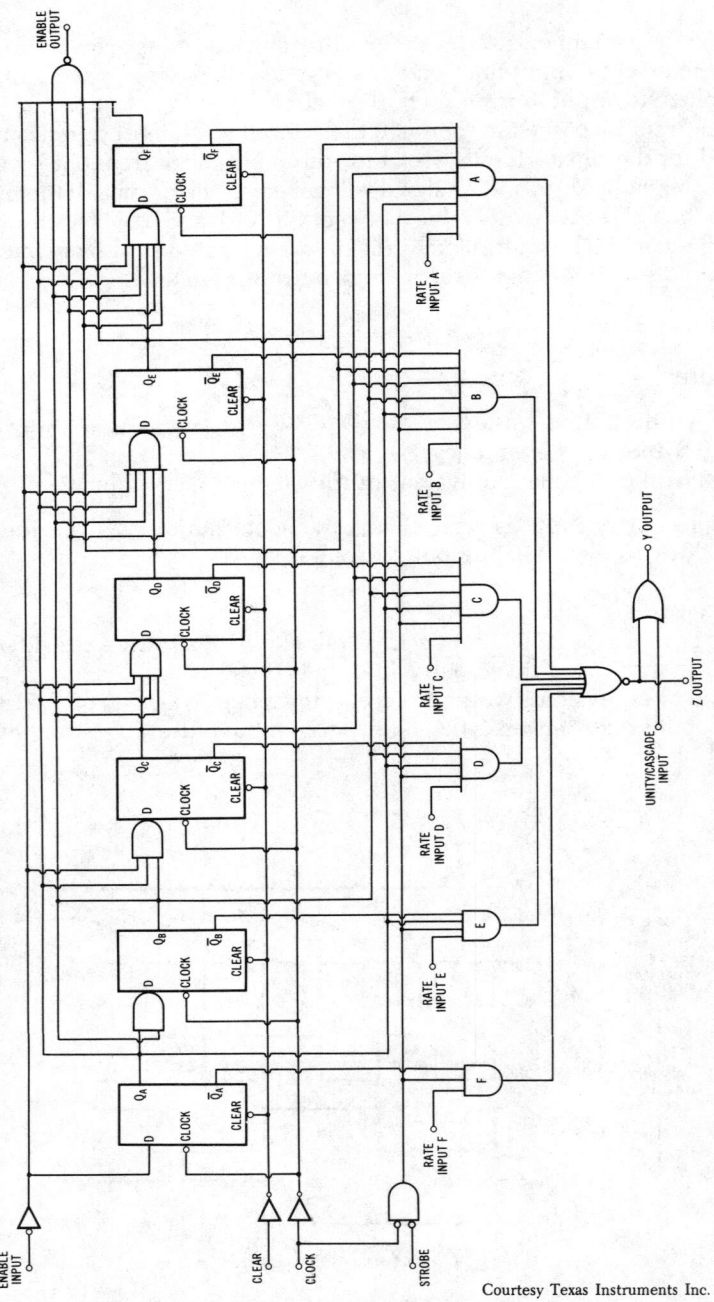

Fig. 9-12. Logic diagram of 7497 binary rate multiplier.

Binary-coded-decimal rate multipliers are also available. The Texas Instruments 74167 and the Motorola 14527 are examples. Each contains four flip-flops and counts in the standard 8-4-2-1 binary code. Otherwise, the arrangement of the circuit is similar to that for the binary rate multiplier. The relationship between the input and output frequencies is expressed by the formula

$$f_o = \frac{Mf_{in}}{10}$$

In the formula, M is the decimal value of the parallel binary input, which is some number between 0 and 9. The bcd rate multiplier provides an output pulse rate based on the bcd number applied to the inputs. For example, if the bcd number 0110 (decimal 6) is applied to the inputs, six output pulses will occur for every ten input pulses.

It is important to note that the output pulses from a rate multiplier are unevenly spaced. The time between successive pulses is not always the same, as it would be in a regular periodic signal. As a result, the output pulse rate is only an average value. We can say that the rate multiplier produces an output frequency that is an average percentage of the input frequency. The result of these unevenly spaced pulses is an effect called *jitter*. In some applications, the jitter may be undesirable, and, therefore, a rate multiplier cannot be used. In other applications, where the unevenness of the output pulse spacing is of no consequence, the binary rate multiplier is an ideal circuit choice because of its ease of programming.

## LSI COUNTERS

As semiconductor technology has improved over the years, it has become possible to put more and more circuitry on a single chip of silicon. First came MSI circuits which are still widely used. Today LSI circuits are available. With LSI techniques, it is possible to put entire digital counter circuits or complete digital systems on a single chip. Microprocessors and microcomputers are an outgrowth of this LSI technology.

An LSI counter is made with MOS rather than bipolar technology. Both p-channel and n-channel MOS techniques are used to make LSI counters. The 7208 and 7216 LSI counters described in a previous chapter are made with CMOS technology. Those LSI counters were designed primarily for frequency measurement. However, there are other types of LSI counters available.

An example of the LSI counters available today is the Mostek 50395 counter. Unlike the 7208 and 7216 discussed in a previous chapter, this device was not designed to be used in general-purpose test equipment for measuring time and frequency. Instead, the design of this counter has been optimized for industrial counting applications. In fact, this particular device is the heart of some commercial industrial counters.

A general block diagram of the 50395 LSI counter is shown in Fig. 9-13. The counter is constructed with p-channel MOS devices. It is contained in a single 40-pin dual–in-line package.

The heart of this device is a six-digit bcd up/down counter circuit. It is capable of counting at rates up to 1 MHz. Up or down counting is selected by an external control pin. A clear input is used to reset the counter. The count input line is buffered with a Schmitt trigger. The count-inhibit line allows the input to be gated by an external signal. A carry output from the most-significant-digit bcd counter allows several of these LSI counters to be cascaded. A zero-detect output is also available. This signal becomes logic 1 when the content of the six-digit bcd counter is zero.

The six bcd counters may be preset. Special counter-load circuitry is incorporated. All six digits may not be preset simultaneously; each of the six bcd counters must be loaded separately. Four bcd input lines are provided for this purpose. These inputs are applied to the load counter multiplexer. These inputs are usually supplied by thumbwheel switches.

Presetting the counter is accomplished in the following manner. The digit counter in the upper right-hand corner of Fig. 9-13 is stepped by an internal oscillator. This digit counter is used in conjunction with the multiplexed display outputs, but it is also used to enable the load counter multiplexer circuit sequentially. The digit counter and a related decoder sequentially enable each counter preset circuit one at a time from the most significant digit to the least significant digit. At each step, the load counter input is checked. If it is a logic 1, the corresponding bcd counter will be preset to the value appearing at the bcd input lines. If the load counter input is a logic 0, the related counter remains unaffected. During the presetting operation, the count input is inhibited to prevent counting from interfering with the loading.

The outputs of the six-digit counter are applied to a six-digit bcd storage register. The number in the counter can be transferred to this six-digit storage register. The storage operation takes place when a logic 1 signal is applied to the store input.

The outputs of the six-digit storage register are applied to an output multiplexer. This multiplexer is operated by the digit counter. The digit counter scans the output multiplexer and enables the bcd outputs from the storage register one at a time so that they appear sequentially at the bcd output lines. In addition to being made available to output pins on the IC, the bcd signal is also applied to a seven-segment decoder. The digit whose bcd output appears at the output of the multiplexer is decoded into standard seven-segment display format. The seven-segment outputs of this LSI counter chip are used to drive external LED displays.

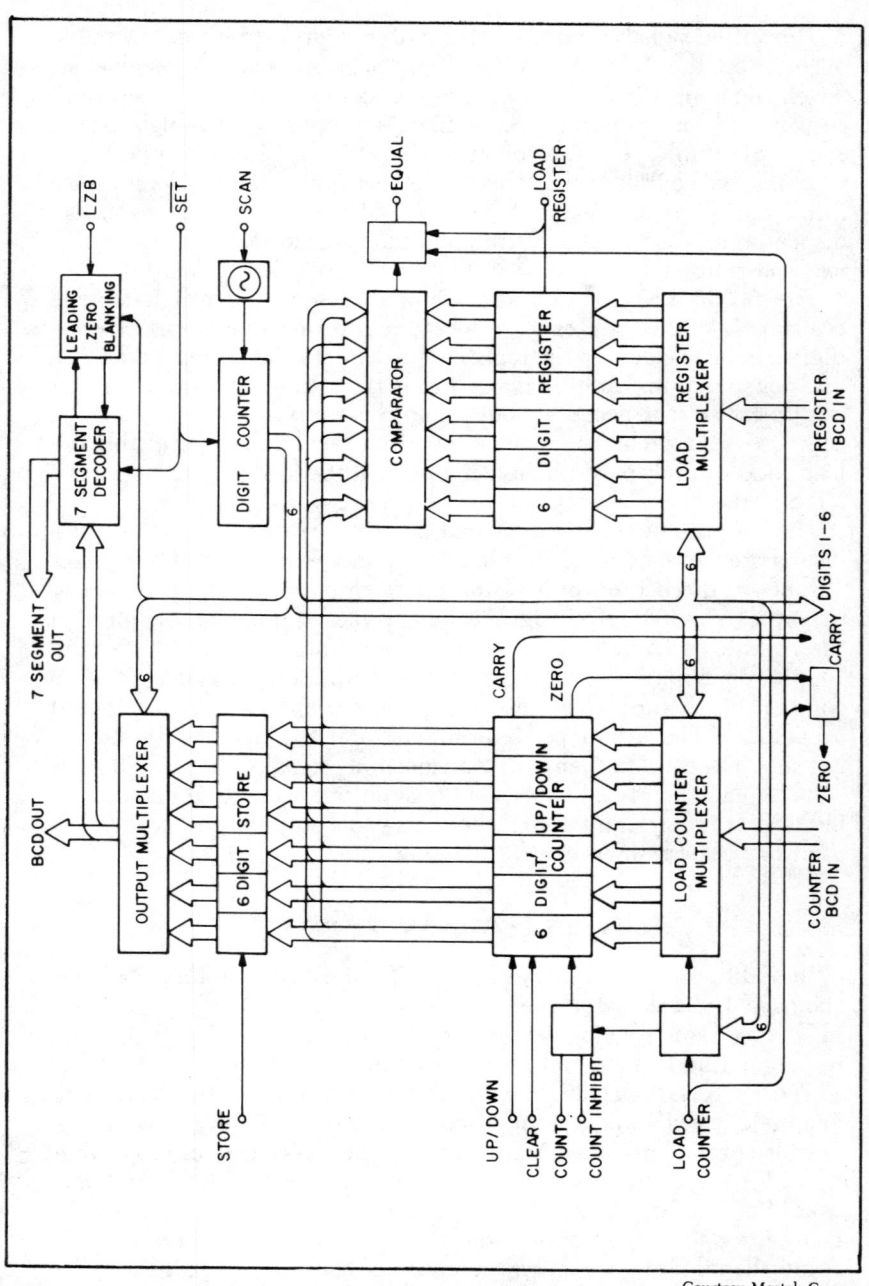

Courtesy Mostek Corp.

Fig. 9-13. Functional diagram of 50395 six-decade counter/display totalizer.

The output signals from the digit counter (digits 1–6) are also available at pins on the IC. Signals on these pins are used to enable six seven-segment display devices sequentially. Each seven-segment display will be enabled one at a time, sequentially, at a high rate of speed. Since the eye cannot follow this, the display gives the appearance of being on continuously. This is the concept behind all multiplex displays. The seven-segment decoder outputs are applied to all six display units. The digit 1-6 strobe outputs are normally connected to external transistors that control the LED displays.

The 50395 LSI counter also contains a second six-digit storage register. Like the counter, this storage register may be preset. Six bcd digits can be loaded into the register one digit at a time. This is done in a manner similar to that used in presetting the counter. The load register multiplexer circuit accepts a four-line bcd input from a source such as a thumbwheel switch. The digit counter and related decoder enable the load register multiplexer a digit at a time. If the load register input is a logic 1, the enabled register will be preset.

The outputs of the six-digit bcd storage register are applied to a comparator circuit. The comparator circuit compares the register outputs to the bcd outputs of the up/down counter. The comparator generates a logic 1 output signal called "Equal" if the number stored in the register is equal to the value in the counter. If the register and counter contents are different, the equal output line is a logical 0. This comparison feature gives the 50395 counter predetermining count capabilities. The desired predetermined count is loaded into the storage register. The counter then counts an incoming signal. When the counter value equals the preset value, an equal output signal is generated. In industrial predetermining counters, this equal output signal is normally used to perform some control function.

## CALCULATORS AS COUNTERS

In addition to the wide variety of MSI and LSI counters available, standard LSI calculator chips can also be used to perform counting functions. This section examines the use of calculator chips in implementing a variety of special counting functions.

The MOS LSI calculator chips widely used to implement the popular handheld calculators can be used to form simple event counters. Because of the low-speed MOS circuitry used in calculator chips, they cannot count at high rates of speed. Most calculator chips have upper count limits of approximately 100 Hz. Despite this limitation, there are many low-speed applications where simple counting functions can be readily handled. Recall that most industrial control applications involve counting speeds that are below 100 Hz. Calculator chips are ideal for such applications.

Despite their counting-speed limitation, calculator chips offer many advantages in the design of special low-speed counters. They can greatly simplify the design and cut costs. To be more specific, consider some of the benefits derived by implementing a counter with a calculator chip:

1. *Multiple Counting Decades*. Most calculator chips have at least six, and more usually eight, decades of counting capability. Chips with 10, 12, or more digits are available. This represents a count capability far beyond that required for most applications.
2. *Internal Display-Drive Capability*. Almost all standard calculator chips have built-in display-driving capability. Using a multiplexed display system, most of these chips can drive fluorescent or LED displays directly. This greatly simplifies the display requirements and reduces the number of external components to a minimum.
3. *Debounce Circuitry*. Calculator chips are operated by keyboards that consist of simple switches. In order to avoid the errors caused by contact bounce, most calculator chips contain internal circuitry that eliminates it.
4. *Counting by Any Radix*. The counting function is implemented on a calculator chip by repeated addition of a constant. To increment a counter by 1, the calculator is initially given the number 1, which is then repeatedly added to accumulate the desired total. The calculator can also be used to accumulate quantities by any other number. For example, the constant 5 can be given to the calculator, and each time the counter is incremented, the count will be incremented by 5 instead of 1. With this capability, a number of unusual or unique applications can be implemented.
5. *Low Package Count and Cost*. Since most of the functions required by a simple event counter are actually contained in the calculator chip, very few external components are required. In most practical applications, a few external devices will be needed for signal conditioning and, of course, the display. Nevertheless, the total component count in such a counter is extremely low. In addition, most calculator chips are extremely low in cost. Because of the high-volume usage, even the most sophisticated calculator chip sells for no more than several dollars. As a result, simple counters can be implemented at very low cost with very few components.
6. *Low Power Consumption*. An MOS calculator chip is basically a low-power device. Most such chips are designed to operate on batteries in a handheld calculator. As a result, calculator counters draw very little current.

The use of a calculator chip as a counter is based on the principle of repeated additions of a constant to accumulate the total. Most modern calculator chips contain what is known as an *autosummation* feature, which means that once a constant is initially entered, subsequent depressions of the add key will result in repeated additions of the constant. The calculator will continually display the result. On a calculator with this autosummation feature, you can demonstrate this to yourself. To do this, first depress the clear key to reset the display at zero. Next, depress the 1 key to enter the constant 1. Then, repeatedly depress the add key. Each time the add key is depressed, the display will be incremented by 1.

Older calculator chips employed a constant operation-counting function. With this type of calculator, the key depression sequence is slightly different. First, depress the clear key to reset the display. Next, the constant 1 is entered with the 1 key. Then the equals key is repeatedly depressed in order to increment the counter. Early calculator chips using what is generally known as *algebraic notation* cannot be used in counter applications.

By using the minus key on the calculator, a down counter can be created. On calculator chips with the autosummation feature, down counting is implemented by using the same procedure described earlier, but substituting the minus key for the plus key. The result is that each time the minus key is depressed, the count will be decremented by one. With both up and down count capability, a wide variety of counting functions can be implemented.

The easiest way to implement a counter with a calculator chip is to start with an entire handheld calculator unit. Why design and build one of these units yourself when you can obtain one for as low as $10? With a few simple additions and modifications, a complete counter can be quickly obtained.

The count input can be received in the form of a contact closure. A push button, toggle switch, slide switch, or set of relay contacts can be connected directly across either the equals or plus key in the calculator, depending on the type of chip used. The regular keyboard can be used to initialize the counter by using the clear and numerical keys. Then each time a contact closure occurs, the calculator will be incremented.

A photocell can also be used as a sensor to increment the counter. A photodiode or resistive photocell can be connected in parallel with the equals or plus key to perform the same function as a switch. For example, in a resistive photocell, the resistance is high when no light is applied. With light applied to the photocell, its resistance drops to a very low value, approximating a short circuit. This will cause the calculator to be incremented.

For applications that require the counting of electronic signals, some form of signal-conditioning circuitry must be used. Electronic pulses can be converted into a contact closure by the use of a Schmitt trigger and a transistor. Each time an electronic pulse occurs, a transistor connected across the plus or equals key can be turned on, thus incrementing the counter.

As indicated earlier, calculators used in this application have an upper frequency limitation. The keyboard-switch debounce circuitry contained on the calculator chip is the primary reason why higher speeds cannot be obtained. Delay circuitry inside the chip ignores bounce noise or actual contact closures for 6 to 30 milliseconds. As a result, maximum counting speed will be somewhere in the range 30 to 150 Hz. This varies widely with the calculator chip, and you should experiment to determine the upper speed limit for your application. If the application should demand a higher counting rate, then standard prescaling circuits with decade counters can be used to extend the frequency range to many kilohertz.

## MICROPROCESSORS AS COUNTERS

Microprocessors can also be used as counters. Because they are general-purpose computers, microprocessors can be programmed to perform the basic counting function and any allied or complementary operation such as display generation. The desirability of using a microprocessor for a counting function depends upon the application. If many other special operations must also be carried out as a part of a larger system that includes counting, the microprocessor may be the best choice. The microprocessor can be programmed to perform calculations and other data manipulations in addition to carrying out a basic counting operation. Like LSI calculator chips, microprocessors when used as counters are typically restricted in their counting speed. Despite this limitation, microprocessors can be used in a number of low-frequency applications.

A typical microprocessor-implemented counter would consist of the CPU chip, a read-only memory (ROM) to store the basic control program, a random-access memory (RAM) where intermediate values and data are stored, and the necessary input/output circuitry. Some type of input-conditioning circuits may be required to make the incoming signal compatible with the microprocessor. If a display is needed to read out the count, then the necessary data-storage registers, decoder/drivers, and display devices are also required.

### Programmed Input/Output

One method of implementing the counting function is to have the signal to be counted applied to one of the microprocessor data-bus lines. In a typical eight-bit microprocessor, you might apply this input signal to the least significant bit of the eight-bit data bus. In order to ensure that the microprocessor recognizes the input signal, it may be necessary to insert a one-shot multivibrator between the input signal and the data-bus line. Each time the input signal occurs, the one-shot multivibrator will generate a fixed-length pulse. The exact length of this pulse will be determined by the frequency of the incoming signal and the speed with which the microprocessor operates.

The microprocessor will be programmed to look repeatedly at the data bus and to watch for the occurrence of an input signal. If the input signal occurs, the accumulator register in the microprocessor will be incremented to indicate the occurrence of an input pulse. The microprocessor will continue to observe the data bus and count the incoming pulses as they occur.

A typical program for such a counting operation is given below. This program is based on the use of instructions for the 8080/8085/Z80 series of microprocessors. (The operation of this microprocessor is certainly beyond the scope of this book. The fundamentals of operation of this processor can be found in other books, such as *The Howard W. Sams Crash Course in Microcomputers*.)

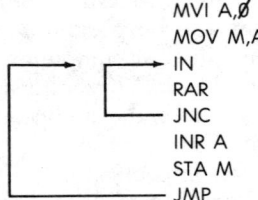

```
        MVI A,Ø
        MOV M,A
     ┌─ IN
   ┌─│─ RAR
   │ └─ JNC
   │    INR A
   │    STA M
   └──  JMP
```

The first instruction, MVI A,Ø, sets the eight-bit accumulator (A) register in the CPU to zero. The second instruction, MOV M,A, transfers the content of the accumulator to a specified memory location, M. This has the effect of clearing, or zeroing, that location, where the count value is to be stored. Next, an input instruction (IN) is executed. This instruction transfers the number on the data bus into the accumulator. In this application, only the count pulse generated by the one-shot on the least-significant bit line is entered. If a count pulse has not occurred, zero will be entered. Once the input instruction puts the data-bus content in the accumulator, an RAR instruction is performed. This shifts the content of the least-significant bit (LSB) of the accumulator into the carry flip-flop, where it may be tested. This test is performed by a JNC instruction, which is executed next. If the LSB was zero, it means that no count pulse occurred. In this case, a jump occurs. The program loops back to the IN instruction. The IN, RAR, JNC instruction loop sequence continues to repeat until an input pulse occurs. When it does, a binary 1 from the one-shot will be loaded into the LSB of the accumulator. After the RAR is executed, the carry flip-flop contains a binary 1. Therefore, no jump occurs. Instead, the next instruction in sequence, INR A, is executed. This instruction increments the accumulator, indicating one input pulse has been counted. The count is then stored away in memory location M by the STA M instruction. A JMP instruction causes the program to loop back around to the IN instruction, where the microprocessor waits for the next input pulse.

This program only illustrates the basic concept of counting with the microprocessor. There are many variations that can be used. For example, if a count capability higher than eight bits (255) is required, then provision must be made to count and store longer binary data words. For example, two consecutive eight-bit words in memory can be used to store the upper and lower halves of a 16-bit word. With 16 bits, a total count capability of 65,535 can be obtained. Additional eight-bit words can be used if the application requires that even larger numbers must be counted.

If displays are to be used to read out the number, then some form of binary-to-bcd conversion must also be performed. This conversion can be implemented with a subroutine in the microprocessor. Another subroutine can be used to address an external storage register for storing the bcd data to be displayed. Each time an input pulse occurs, the accumulator register will be incremented and the number stored in memory. The microprocessor will then perform the binary-to-bcd conversion and output the updated number to the display. This process will continue as input pulses occur.

## Interrupt Counting

Another method of implementing the count function is to use the interrupt line of the microprocessor. Instead of applying the input signal to the data bus, which is constantly being monitored by the CPU, the count pulse is applied to the interrupt line. Each time an interrupt occurs, the microprocessor automatically branches to a subroutine that performs the count operation. The advantage of the interrupt technique over the programmed i/o method is that the microprocessor is left free to perform other operations between count pulses. In some applications where the count pulses occur at a very slow rate, there is no reason why the microprocessor should wait for input pulses if it can be doing something else. The interrupt counting method is also faster.

As in the programmed i/o application, a one-shot multivibrator may be connected between the incoming pulses and the interrupt line. When the interrupt occurs, the microprocessor completes the execution of any instruction currently in process and then automatically branches to a subroutine that services the interrupt and implements the counting function. Before branching to the count subroutine, the microprocessor may store the data in the accumulator or other registers prior to implementing the count function. This will allow any important data to be saved so that the program in process can be restarted once the count has been tallied. Once the microprocessor tallies the input count, it will return to the program in process, and the regular program will continue to be executed until another input pulse occurs.

An example of an interrupt count program using the 6800 microprocessor is shown below. These instructions effectively implement a counting routine similar to that illustrated in the previous example.

```
LDAA M
INCA
STAA M
RET
```

When an interrupt occurs, the 6800 microprocessor completes the execution of any instruction in progress and then automatically stores all register contents in the stack. This prevents loss of any data or intermediate results while the CPU executes the count subroutine. After this, the 6800 branches to the subroutine. The first instruction, LDAA M, loads the accumulator with the content of memory location M. This is the location designated to store the count. This location must have been initialized to zero at a previous time.

The next instruction increments the accumulator by one. Then, the accumulator content is stored in memory location M by the STAA M instruction. Finally, a RET (return) instruction is executed. This causes all previous register contents and CPU status to be restored. The program in progress when the count interrupt occurred resumes and will continue until another interrupt occurs.

Binary-to-bcd subroutines, display output routines, calculation functions, and other operations can also be implemented to enhance the counting function if the application demands it.

Of the two microprocessor counting techniques, the interrupt method is preferred. It leaves the microprocessor free for other operations and is typically faster. The actual speed of counting in both of these methods should be several thousand counts per second if a high-speed microprocessor is used. Keep in mind that the actual counting rate will depend on the clock frequency of the microprocessor, the instruction cycle time, the length and complexity of the program, and the exact counting technique used. Another important factor in counting speed is the other activity that must be performed by the microprocessor. Binary-to-bcd conversion, display subroutines, and calculation subroutines require a finite amount of time to execute. If these operations are to take place between count pulses, then naturally they will affect the counting speed. It is not possible to give specific numbers because the range is far too wide and depends on the specific application.

## On-Board Counters

Counting operations are so common in digital systems that nearly every application requires them. Since microprocessors are more widely used every day in implementing digital systems, it makes sense to give them a special count capability. This has been done in some of the newer single-chip microcomputers. These devices contain not only the normal CPU but also the RAM, ROM, and i/o circuits on a single chip. In addition, most of these devices incorporate at least one and sometimes more 8- or 16-bit counters. Typical devices are the Intel MCS-48 family (8048, 8021, etc.), the Mostek/Fairchild 3870 series, the Zilog Z8, the Motorola 6801, and others. These on-chip counters are often called timers and come in many different configurations.

Associated with these on-board counters are a series of instructions that can load the counter from memory or another CPU register or transfer the counter content to memory or other registers. Other instructions start or stop the count. Most of the counters generate an overflow that can be tested to create decision-making program jumps. Some counters also generate interrupts. An input pin is provided so that external events may be counted.

The advantages of an on-board counter are that special count subroutines like those described earlier are simply not needed. In addition, higher counting speeds can be obtained. On-board counters also eliminate the need for external counters and the related interface circuitry.

Some of the applications of on-board counters include event counting, time delay, pulse-width generation and measurement, frequency measurement, and timekeeping. Timing sequences usually generated by programmed loops can also be eliminated in many applications. Almost all timing and counting operations previously programmed can now be provided by the on-board counters.

## MICROPROCESSOR PERIPHERAL COUNTER/TIMERS

While many counting operations can be adequately handled by count subroutines or on-board counters in microprocessors, there are some applications that have special requirements that cannot be handled. These may involve higher counting speeds, multiple counting functions, or some other unusual conditions. In many of these cases, regular MSI and/or LSI counter circuits can be used. When such devices are used, interface circuitry must also be employed to interconnect the counter circuits and the microprocessor. The result is often a complex and costly solution to the problem.

An alternative to these external counting circuits is a new class of MOS LSI counter/timer device designed exclusively as a peripheral chip for microprocessors. These ICs connect directly to the microprocessor data and address buses and require no special interfacing. These chips are fully programmable and can be used to implement virtually any timing or counting function. We will take a look at three of these units.

## Am9513

The Am9513 is called a system timing controller. A general block diagram of this device is shown in Fig. 9-14. It consists of a clock oscillator, a 16-bit frequency scaler, five counter logic groups, and all of the command, status, mode, and interface control circuitry. The heart of each counter logic group is a 16-bit counter that can be programmed to count up or down in binary or bcd. The maximum count rate is 7 MHz. These counters may be cascaded to increase the count capability. The configuration of each counter logic group is programmed by a 16-bit mode word stored in the mode registers. Besides up, down, bcd, or binary counting, the polarity of the input count pulse and up to 16 different input sources can be selected.

Fig. 9-15 shows the counter logic groups in more detail. In addition to the main 16-bit counter, there are load and hold registers. These, like the main counter, can be loaded from the microprocessor or have their contents transferred to the microprocessor. The load register is used to preset the counter. The hold register is used to save count values. The hold register may also be used as a second load register. Note that counter logic groups 1 and 2 also have a comparator and an alarm register (Fig. 9-15A). The comparator compares the contents of the alarm register and the main counter and generates an output pulse if the two are equal.

The 9513 system timing controller operates under software control from the microprocessor. Command and mode words are transmitted over the data bus to set up the desired configuration and mode of operation. Data to be stored in the counters or registers is sent via the CPU over the data bus. The counter contents can be read under program control. Counter gating can be controlled by external hardware-generated signals or a software command.

Some of the potential applications of this device are:

Frequency measurement
Time-interval or period measurement
Pulse-width generation
Frequency synthesis
Predetermining count decisions
Time delays
Pulse sequences
Event counting

## MC6840

Another peripheral counter chip is the Motorola MC6840 programmable timer module. A block diagram of the MC6840 is shown in Fig. 9-16. It consists of three 16-bit down counters, which may be decremented by external signals or the internal clock. The signals to be

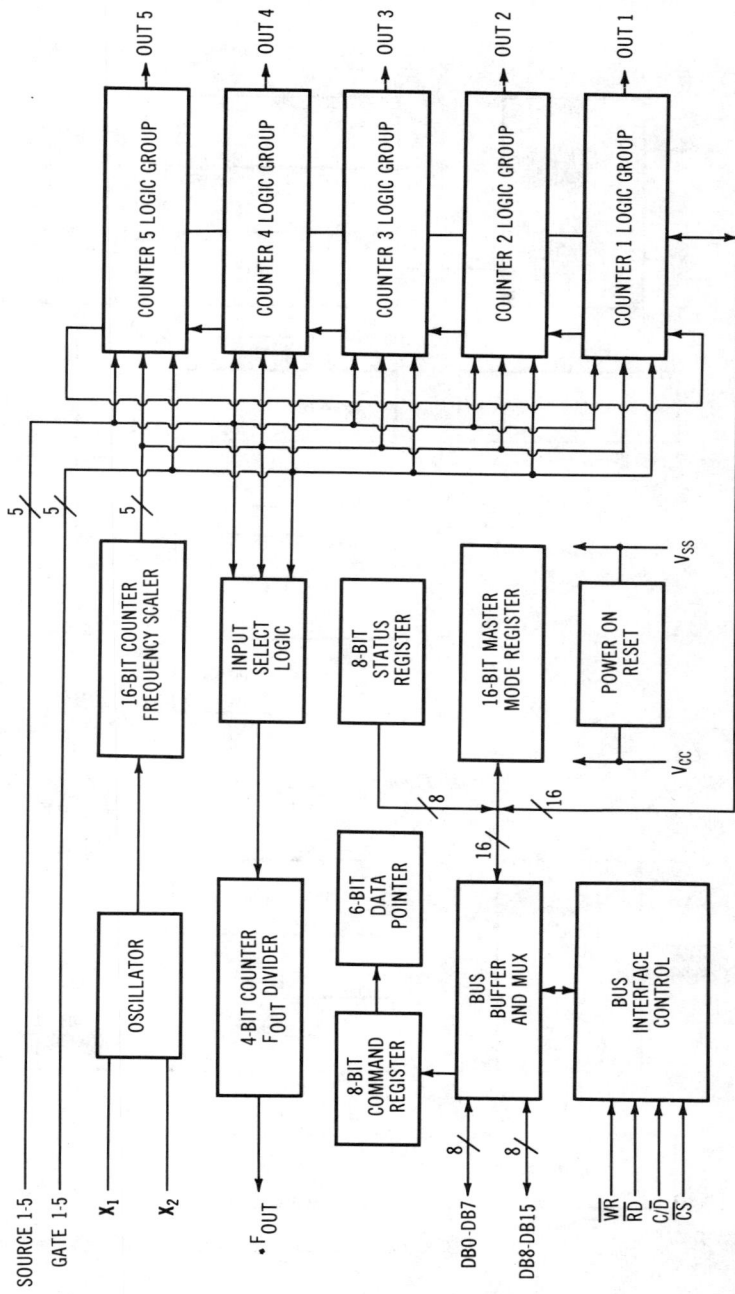

Fig. 9-14. Block diagram of Am9513 MOS LSI system timing controller.

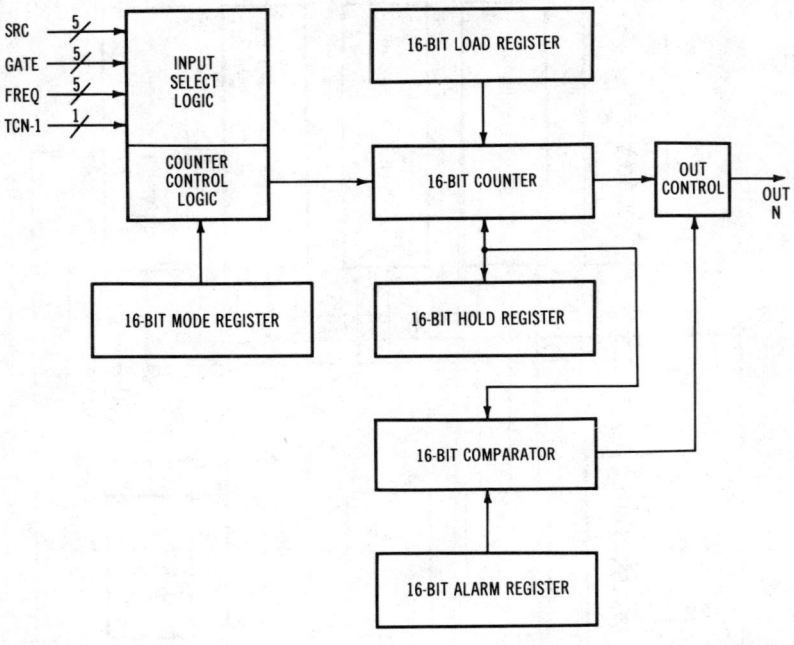

*(A) Groups 1 and 2.*

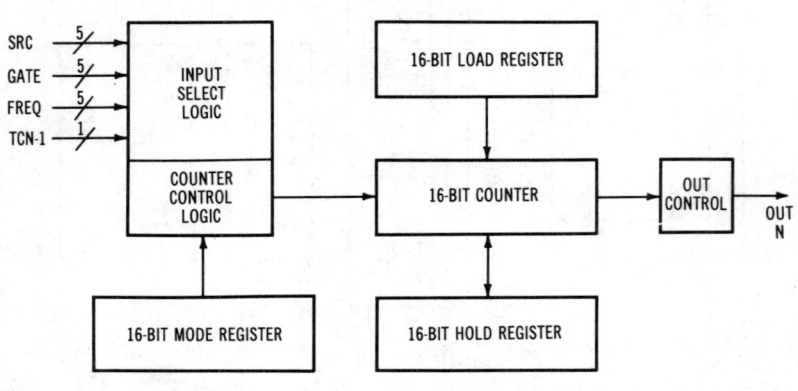

*(B) Groups 3, 4, and 5.*

Courtesy Advanced Micro Devices, Inc.

Fig. 9-15. Details of counter logic groups.

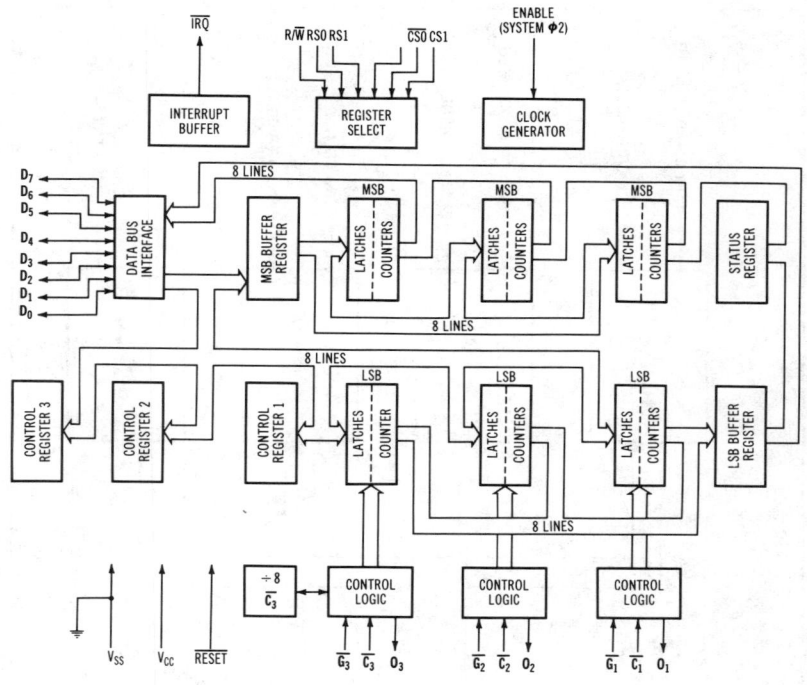

Fig. 9-16. Block diagram of MC6840 programmable timer module.

counted are gated by external pulses. Each counter has a 16-bit storage register associated with it. These latches contain the count that is to be preset into the counters. The counters may be used to cause system interrupts and/or to generate output pulses.

Three eight-bit control registers are used to set up the counter modes. These are loaded under program control via the data bus. The main operating modes are continuous, single-shot, frequency comparison, and pulse-width comparison. In the continuous mode, the counter generates a square wave or variable-duty-cycle pulse train whose intervals are determined by the counter preset values and the clock frequency.

In the one-shot mode, the counter simply generates a single output pulse whose width is a function of the counter preset value and the clock frequency.

In the frequency and pulse-width comparison modes, the counters are decremented by input pulses for a duration determined by the external gating signals. If the counter is decremented to zero before the gate interval ends, an interrupt is generated. With this arrangement,

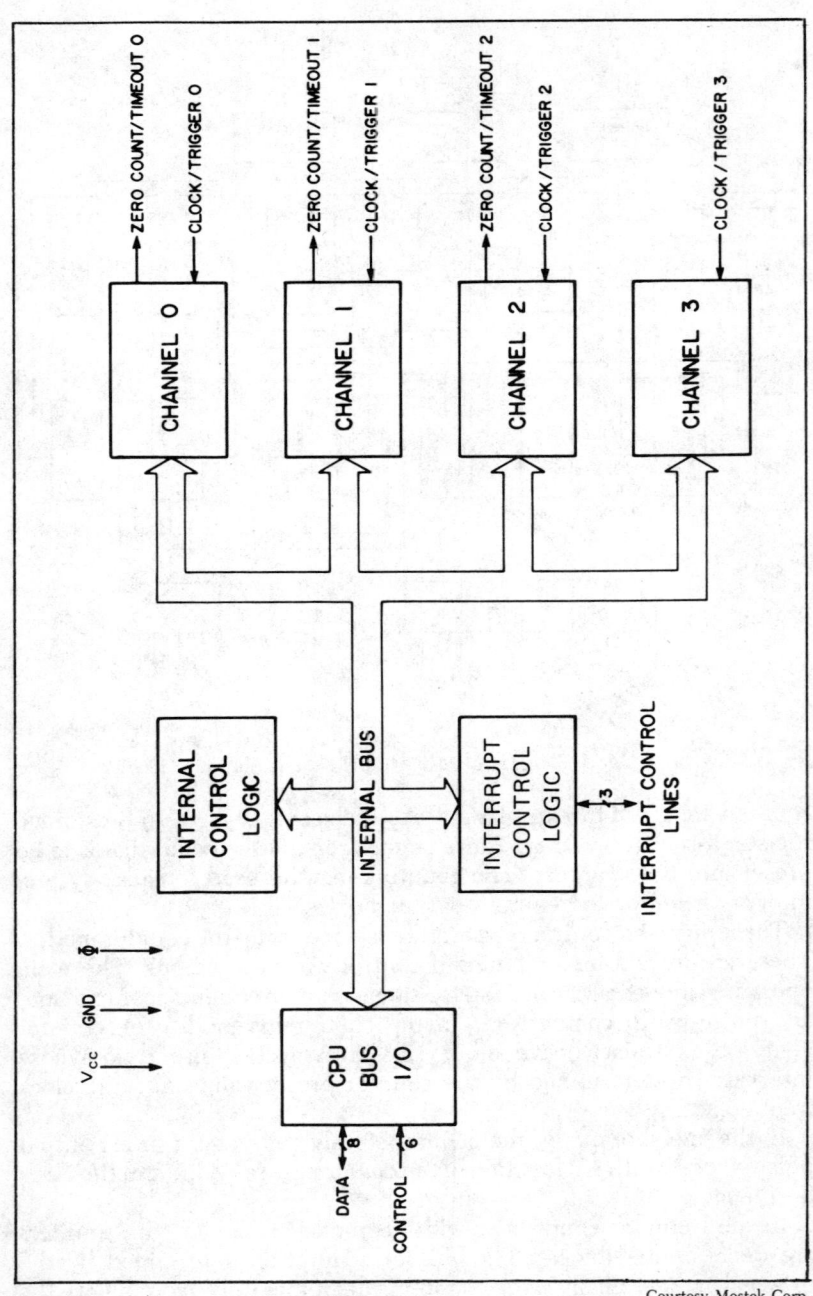

Fig. 9-17. Block diagram of Z80-CTC LSI counter IC.

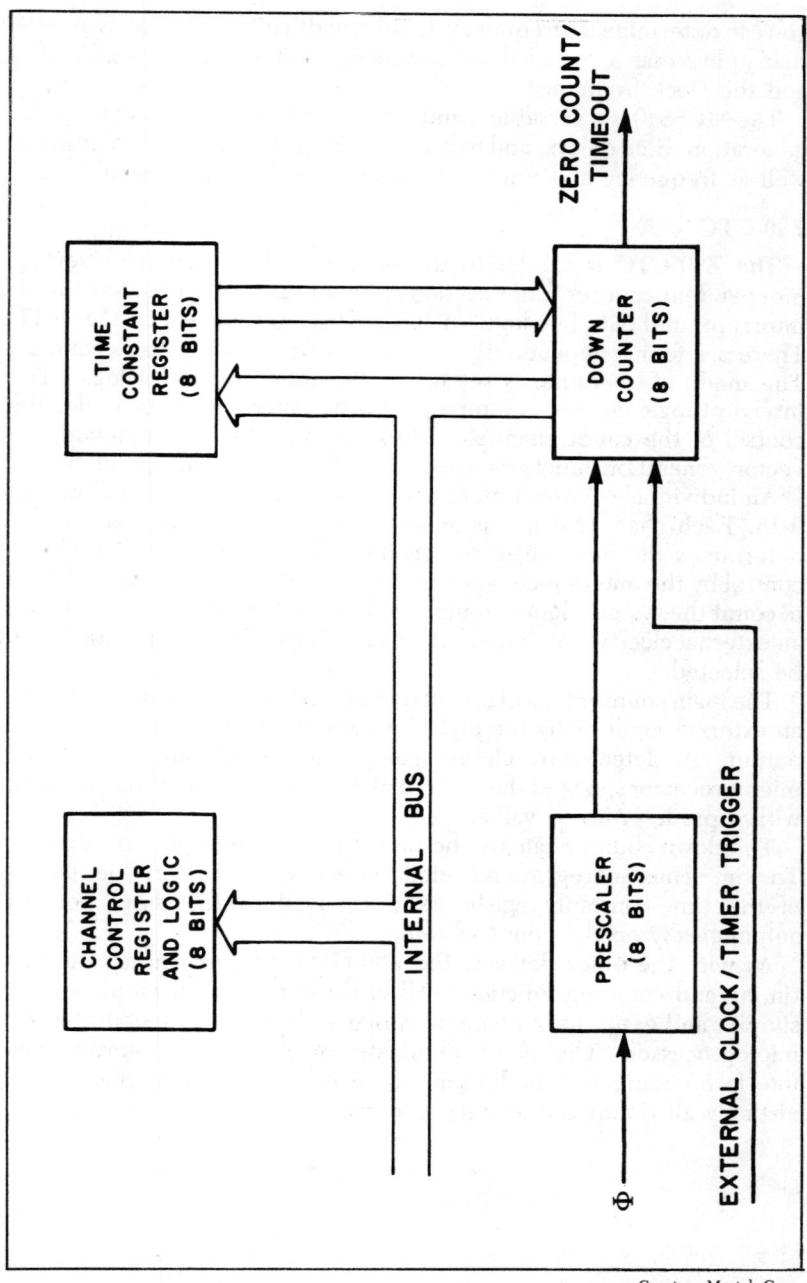

Fig. 9-18. Details of counter channel.

Courtesy Mostek Corp.

you can determine if a frequency or pulse width (time interval) is greater than or less than a given value that depends on the counter preset value and the clock frequency.

The MC6840 is a versatile counter/timer device. It is useful for pulse generation, time delays, and frequency synthesis, but event counting as well as frequency and time measurements can be accommodated.

## Z80-CTC

The Z80-CTC is similar to the others in that it contains several independent counter/timer sections plus all the related mode-control, interrupt, and data-bus logic. A block diagram is shown in Fig. 9-17. There are four independently programmable timer/counter channels. The mode of operation is set up by the internal control logic. The interrupt logic generates interrupts for the microprocessor under the control of the count channels. Interrupt priority determination and vector generation functions are also performed by this circuitry.

An individual counter/timer channel is shown in more detail in Fig. 9-18. Each channel contains an eight-bit mode-control register that determines the function of the channel. It is loaded under program control by the microprocessor. For example, the counters can be set up to count the system clock through a $\div 16$ or $\div 256$ prescaler, or to count an external clock input. Triggering by a rising or falling input pulse can be selected.

The main counter is an eight-bit down counter that is decremented by an external input or by the eight-bit prescaler from its $\div 16$ or $\div 256$ output. A detect-zero circuit generates the output signal. The microprocessor can read the content of this counter at any time or load it with a predetermined value.

The down counter can also be preset by the time-constant register. The time-constant register is loaded by the microprocessor. The content of the time-constant register is transferred to the down counter automatically on the count of zero.

As with the other devices, the Z80-CTC can perform a variety of timing and counting functions. All of these peripheral chips make it simpler and easier to implement timing and counting functions with a microprocessor. The chips eliminate many special counting and interface circuits and at the same time relieve the microprocessor of virtually all timing and counting software overhead.

# Index

## A

Accuracy, 43, 63, 102-103
Acquisition period, 103
Active probes, 118
Add/subtract counters, 166
Address(es), 229
  generation, 227-231
Adjustments, counter, 137-140
Agc, 103
Algebraic notation, 246
Am tolerance, 103-104
Amplifier, input, 23-24
Amplitude discrimination, 104
Antenna, 118-119, 137
Arming circuits, 132
ASCII code, 129
Asynchronous, synchronous versus, 218-222
Attenuator, 122-124
  input, 23
Auto-ranging, 44
Automatic
  gain control, 103
  heterodyne conversion system, 84-88
  transfer-oscillator converter, 95-97
  trigger-level setting, 158
Autosummation, 245-246
Averaging, period, 49

## B

Bandwidth, 61-62
Bcd, 11
  /binary conversions, 227
  counter, 11-13
    applications of, 226-232
  interface, 127-129

Bench counter, general-purpose, 198-205
  circuit operation, 199-205
  specifications, 198-199
Binary
  /bcd conversions, 227
  coded decimal, 11
  numbers, 11
  counter, 8-11
    applications of, 226-232
Blocks, 229
BNC connector, 119-120
Bounce, contact, 135
Bytes, 129, 227

## C

Cable(s), 108-128
  coaxial, 109
Calculators as counters, 244-247
Capacitance
  cable, 110-113
  input, 60
Capacitor
  compensating, 116
  trimmer, 115
Characteristic impedance, 123
Circuits, counter, 7-13
Clock, digital, 169
Clocked counter, 218
CMOS counters, 218
Coil, sniffer, 136-137
Communications counter, LSI, 187-198
  circuit operation, 189-198
  specifications, 188-189
Comparator
  circuit, 58-59
  digital, 231

259

Compensating capacitor, 116
Complement
  output, 8
  tens, 231-232
Complex signal measurements, 152-155
Components, counter, 20-35
Computing counters, 106-107, 156-161
  advantages and benefits of, 160-161
  functions of, 158-160
Connecting the signal source, 134-137
Connector(s), 108-128, 136
  attenuator, 122-124
  BNC, 119-120
  converter, 121
  D, 129
  fuse, 122
  T, 120-121
  terminator, 121-122
Contact bounce, 155
Control circuits, 29-32
Conversion(s)
  bcd/binary, 227
  down, 76
    heterodyne, 79-88
    transfer-oscillator, 88-97
  heterodyne, manual, 81-84
  system, heterodyne, automatic, 84-88
Converter
  connector, 121
  digital-to-analog, 87
  harmonic-heterodyne, 97-100
  transfer-oscillator, automatic, 95-97
Counter(s)
  add/subtract, 166
  adjustments, 137-140
  bcd, 11-13
    applications of, 226-232
  bench, general-purpose, 198-205
    circuit operation, 199-205
    specifications, 198-199
  binary, 8-11
    applications of, 226-232
  calculators as, 244-247
  circuits, 7-13
  clocked, 218
  CMOS, 218
  communications, LSI, 187-198
    circuit operation, 189-198
    specifications, 188-189
  components, 20-35
  computing, 106-107, 156-161
    advantages and benefits of, 160-161
    functions of, 158-160
  decade, 13
  dividers, as, 225

Counter(s)—cont
  ECL, 218
  electromechanical, 162
  electronic, 167-172
  frequency-multiplier, 107
  grounding of, 117
  high-frequency, specifications of, 100-105
  industrial, 18-19, 161-177
    applications of, 174-177
  integrated-circuit, 33-35
  lineal, 165-166
  low-frequency, 105-107
  LSI, 35, 241-244
    handheld, 179-185
      circuit operation, 181-185
      specifications, 179-181
  measuring, 165-166
  mechanical, 162-167
    advantages of, 167
  microprocessors as, 247-251
  MSI, 217-218
  on-board, 250-251
  operational modes, 35-54
    frequency-measurement, 35-44
    period, 46-50
    ratio, 44-46
    time-interval measurement, 52-54
    totalizing, 50-52
  predetermining, 166
  printing, 167
  ripple, 218-219
  section, 25-26
  stroke, 165
  synchronous, 219-222
  test instrument, electronic, 14-18
  /timers
    peripheral, microprocessor, 251-258
    universal, 205-216
      counter operation, 208-216
      specifications and features, 205-208
  totalizing, electrical, 165
  TTL, 218
  up/down, 166, 223
Counting
  interrupt, 249-251
  radix, by any, 245
Coupling circuit, ac/dc, 22
Crystal, 33
  oscillators, 64
    temperature-compensated, 66
  oven, 66
  temperature effects on, 65-66
Cutoff frequency, 61-62
Cycle, duty, 14, 146-147

## D

D connector, 129
Dac, 87
dBm, 102
Debounce circuitry, 245
Decade(s)
 counters, 13
 counting, multiple, 245
Decoding
 glitches, 219
 spikes, 219
Delay, propagation, 147-149, 219, 220
 average, 148-149
Digital
 clock, 169
 frequency display, 231-232
 -to-analog converter, 87
Digits, number of, 101
Diodes, step recovery, 82-83
Discrete multiplier, 237
Discrimination, amplitude, 104
Disk, floppy-, storage system, 229-231
Display
 -drive capability, internal, 245
 frequency, digital, 231-232
 multiplexed, 29
 section, 26-29
Dividers
 counters as, 225
 frequency, 226
Down conversion, 76
 heterodyne, 79-88
 transfer-oscillator, 88-97
Drift, long-term, 66-67
Drive capability, display, internal, 245
Duty cycle, 144, 146-147
Dynamic range, 63, 102

## E

ECL counters, 218
Electrical totalizing counters, 163
Electromechanical counters, 162
Electronic counter(s), 167-172
 test instrument, 14-18
Error, 69
 ±1-count, 68-71, 106
 quantization, 69
 sources, 56, 68-74
  miscellaneous, 73-74
 trigger, 71-73

## F

Fall time, 144, 147
FET follower, 118

Filters
 imput, 73
 yig, 86
Flat line, 123
Flip-flop operation, 8-9
Floppy-disk storage system, 229-231
Fm tolerance, 103-104
Follower, FET, 118
Frequency
 cutoff, 61-62
 display, digital, 231-232
 dividers, 226
 measurement, 134-141, 152-155
  mode, 35-44
 measuring devices, early, 17-18
 meter, heterodyne, 17-18
 multiplier counters, 107
 range, 60-62, 101
 ratio measurements, 140-141
 synthesizer, 98
 upper limit on, 75
Fuse, connector, 122

## G

Gate, main, 25
General-purpose
 bench counter, 198-205
  circuit operation, 199-205
  specifications, 198-199
 interface bus, 131
Generation, address, 227-231
Generator(s)
 harmonic, 81, 90-91
 pulse, rotary, 174
Glitches
 decoding, 219
 noise, 153-154
GPIB, 131
Grounding, 135-136
 counter, 117
 lead, 117

## H

Handheld LSI counter, 179-185
 circuit operation, 181-185
 specifications, 179-181
Harmonic
 generator, 81, 90-91
 heterodyne converter, 97-100
Hertz, 35
Heterodyne
 conversion
  manual, 81-84
  system, automatic, 84-88

Heterodyne—cont
  converter, harmonic, 97-100
  down conversion, 79-88
  frequency meter, 17-18
High-frequency
  counters, specifications of, 100-105
  probes, passive, 110-117
Hold-off, TI, 73
Hysteresis, 58, 59
  window, 58

## I

IEEE-488 interface, 131, 216
Impedance
  characteristic, 123
  input, 59-60, 102
  surge, 123
Industrial counters, 18-19, 161-177
  applications of, 174-177
Input
  circuit, 22-24
  impedance, 59-60, 102
  /output, programmed, 247-249
  voltage, maximum, 63
Integrated circuit
  Am9513, 252
  counters, 33-35
  MC6840, 252-258
  XR-2242, 233
  Z8, 251
  Z80-CTC, 258
  74LS90, 203, 204
  3870 series, 251
  6801, 251
  7207, 185
  7208, 182-185
  7216D, 189-192
  7492, 204
  7497, 237-239
  8021, 251
  8048, 216, 251
  8240, 233-237
  8250, 233-237
  9368, 203-204
  50395, 241-244
  68488, 216
  74167, 241
  74196, 197, 203
Integration, medium-scale, 217
Intelligent instruments, 157
Interconnections, 133
Interface(s), 126-131
  bcd, 127-129
  bus, general purpose, 131

Interface(s)—cont
  IEEE-488, 131
  RS-232, 129
Interrupt counting, 249-251
Inversion, phase, 151

## J

Jitter, 241

## L

Large-scale integration, 35
Lead, grounding, 117
Limiter, voltage, 23
Line, transmission, 123
  flat, 123
Lineal counters, 165-166
Listeners, 131
Long-term
  drift, 66-67
  stability, 67
Loop, pickup, 119
Low-frequency counters, 105-107
LSI, 35
  communications counter, 187-198
    circuit operation, 189-198
    specifications, 188-189
  counter(s), 35, 241-244

## M

Manual
  heterodyne conversion, 81-84
  transfer oscillator, 89-95
Masking, trigger, 73
Measurement(s)
  complex signal, 152-155
  frequency, 152-155
    ratio, 140-141
  practices, good, 133-134
  principles, basic, 132-134
  pulse, 143-147
  speed, 103
  time, 141-152, 155
Measuring counters, 165-166
Mechanical counters, 162-167
  advantages of, 167
Medium-scale integration, 217
Microhertz, 107
Microprocessor(s), 99, 157
  as counters, 247-251
  peripheral counter/timers, 251-258
Mismatch, 123
Mixers, 80
Modes, counter operational, 35-54
  frequency-measurement, 35-44

Modes, counter operational—cont
  period, 46-50
  ratio, 44-46
  time-interval measurement, 52-54
  totalizing, 50-52
MSI, 217
  counters, 217-218
Multiplexed display, 29, 30-32
Multiplier
  discrete, 237
  rate, 237-241

### N

Noise, 71-72
Noncoherence, 69
Nonlinear operation, 63
Notation, algebraic, 246
Numbers, binary coded decimal, 11

### O

On-board counters, 250-251
Oscillators
  crystal, 33, 64
    temperature-compensated, 33, 66
  local, frequency of, 81
  time-base, 64
  transfer
    converter, automatic, 95-97
    manual, 89-95
  variable-frequency, 88
  voltage-controlled, 90
Oscilloscope, 133-134
Oven, crystal, 66
Over-ranging, 44

### P

Package
  cost, 245
  count, 245
Parts per million, 66, 67
Passive probes, high-frequency, 110-117
Performance specifications, 56-68
Period, 144
  acquisition, 103
  averaging, 49
    mode, 142-143
  measurement, 141-142
  mode, 46-50, 106
Peripheral counter/timers, microprocessor, 251-258
Phase
  inversion, 151
  locked loop, 90, 92-93, 95, 107
  shift, 149-152
Photocells, 173, 246

Pickup(s)
  loop, 119
  rf, 118-119
Pll, 90, 92-93
Power
  consumption, 245
  supply voltage, effect of on stability, 67
Preamplifiers, 124-125
Precision, 43
Predetermining counters, 166
Prescalers, 125-126
Prescaling, 76-79
Preset, 223
Printer, 127
Printing counters, 167
Probe(s), 108-124
  active, 118
  attenuator, 113-117
  clip-on, 135
  passive, high-frequency, 110-117
Programmed input/output, 247-249
Propagation delay, 147-149, 219, 220
  average, 148-149
Proximity switches, 173-174
Pulse
  characteristics, 144
  generators, rotary, 174
  measurements, 143-147
  spacing, 144, 146
  width, 144, 145-146

### R

Radian, 151
Radix, counting by any, 245
Rate multiplier, 237-241
Ratio mode, 44-46
Reed switches, 173
Register, storage, 28
Relays, 173
Reset, 222-223
  state, 8
Resolution, 40-42, 43, 62
  loss in prescaling, 78
Rf pickups, 118-119
Ripple counters, 218-219
Rise time, 144, 147
Rotary pulse generators, 174
RS-232 interface, 129

### S

Sampler, 95
Schmitt trigger, 58
Sectors, 229
Self-test, 158
Sensitivity, 56-59, 102

Sensors, 172-174
Set state, 8
Shift, phase, 149-152
Short-term stability, 67
Signal
　complex, measurements of, 152-155
　source, connecting, 134-137
　-to-noise ratio, 104
Significant figures, 43
Slope, 143-145
Sniffer coil, 136-137
Snr, 104
Spacing, pulse, 144, 146
Specifications, 55
　comparison of, 105
　high-frequency counters, 100-105
　performance, 56-68
Speed, measurement, 103
Stability
　long-term, 67
　short-term, 67
　voltage effect on, 67
Step recovery diodes, 82-83
Storage system, floppy-disk, 229-231
Stroke counters, 165
Surge impedance, 123
Switches, 172
　proximity, 173-174
　reed, 173
Synchronous
　asynchronous, versus, 218-222
　counters, 219-222
Synthesizer, frequency, 98

## T

T connector, 120-121
Talkers, 131
Tcxo, 33, 66
Temperature
　-compensated crystal oscillator, 66
　crystal, 65-66
Tens complement, 231-232
Termination, 122
Terminator, 121-122
TI hold-off, 73, 152
Time
　base, 33
　　characteristics 63-68
　　extender, 91-93
　　signals, 64-65
　fall, 144, 147
　interval
　　hold-off, 73, 152
　　measurement mode, 52-54

Time—cont
　measurements, 141-152, 155
　rise, 144, 147
Timer(s), 233-237
　electronic, 176-177
Tolerance
　am, 103-104
　fm, 103-104
Totalizing
　counters, electrical, 165
　mode, 50-52
Transfer oscillator
　converter, automatic, 95-97
　down conversion, 88-97
　manual, 89-95
Transmission line, 123
　flat, 123
Trigger
　circuit, 24
　error, 71-73
　level setting, automatic, 158
　masking, 73
Triggering level, 145
Trimmer capacitor, 115
TTL counter, 218

## U

Universal counter/timer, 205-216
　counter operation, 208-216
　specifications and features, 205-208
Up/down counters, 166, 223-225

## V

Vco, 90
Vfo, 88, 90
Voltage
　controlled oscillator, 90
　input, maximum, 63
　limiter, 23
　power-supply, effect of on stability, 67
　standing-wave ratio, 122, 123
Vswr, 122, 123

## W

Wavemeters, 17
Width, pulse, 144, 145-146
Window, hysteresis, 58

## Y

Yig filters, 86

## Z

Zero beat, 89